Praise for *Scientists Greater than Einstein*

These remarkable stories of human achievement on a truly grand scale are told with great sensitivity, yet fine precision. These are the gritty details behind some of the greatest medical advances in human history. Woodward's story goes deep, showing us the personal lives of the players involved and where they were when the eureka moment came, then follows each researcher's idea through to its full objective impact. The hard core measure of human lives saved is the final irrevocable statement of what can be accomplished in a human being's lifetime.

—Jeff Wicker, MD

This engaging book is a fascinating read chronicling the genesis of scientific curiosity, the travails and myriad failures, and, ultimately, the triumphs of those researchers responsible for promoting the advancement and survival of mankind. It should promote bountiful discussion and would be an excellent library addition for students of science of all ages who seek to understand the price and impact of scientific progress.

—Susan R. Campbell Ph.D.
Chair, Chemistry Department
Georgetown College

Scientists
Greater
than
Einstein

The Biggest Lifesavers
of the Twentieth Century

by Billy Woodward

Quill
Driver
Books

Fresno, California

Printed in the United States of America.

Published by
Quill Driver Books, an imprint of Linden Publishing
2006 S. Mary, Fresno, CA 93721
559-233-6633 / 800-345-4447
QuillDriverBooks.com

Quill Driver Books may be purchased for educational, fund-raising, business
or promotional use. Please contact Special Markets, Quill Driver Books,
at the above address or phone numbers.

Quill Driver Books Project Cadre:
Christine Hernandez, Maura J. Zimmer, John David Marion,
Sylvia Coates, Stephen Blake Mettee, Kent Sorsky

135798642
ISBN 978-1884956-87-4 • 1-884956-87-4

To order a copy of this book, please call
1-800-345-4447.

Edited by Gabriel Popkin

Library of Congress Cataloging-in-Publication Data

Woodward, Billy, 1956-
  Scientists greater than Einstein : the biggest lifesavers of the twentieth cen-
tury / by Billy Woodward, Joel N. Shurkin, Debra Gordon.
      p. ; cm.
  Includes bibliographical references and index.
  ISBN-13: 978-1-884956-87-4 (hardcover)
  ISBN-10: 1-884956-87-4 (hardcover)
  1. Medical scientists--Biography. 2. Medicine--Research--History--20th
century. 3. Medical innovations--History--20th century. I. Shurkin, Joel
N., 1938- II. Gordon, Debra L., 1962- III. Title.
  [DNLM: 1. Medicine--Biography. 2. History, 20th Century. 3. Research-
-history. WZ 112 W899s 2008]
  R149.W596 2008
  610.92'2--dc22

2008030163

Dedicated to the memory of Gertrude Elion and George Hitchings, for their development of Allopurinol, which has so greatly enhanced my life.

—Billy Woodward

Written by   Billy Woodward
                      Joel Shurkin
                      Debra Gordon

Edited by   Gabriel Popkin

Biostatistics by   Amy R. Pearce, Ph.D.

Contents

Acknowledgments

If the only prayer you said in your whole life was 'thank you,'
that would suffice.

—Thirteenth century German theologian Meister Eckhart

IT WAS a delight conversing with Al Sommer, Akira Endo, Bill Foege, and David Nalin and I greatly appreciate the time and effort they took to share their stories and photographs, and to read and make corrections to their respective chapters.

The book would not have been completed with any degree of professionalism without Gabriel Popkin. His help in researching the subject from the germ of the idea to editing the last word was indispensible. Most importantly, he insisted that the science be presented correctly.

I am grateful for the writing experience and enthusiasm Joel Shurkin and Debra Gordon brought to the book. I wish to thank Amy Pearce for her diligent work in calculating the numbers of lives each scientist saved and Hiroko Hollis for translating three Japanese books for Akira Endo's chapter. I also would like to thank Chuck Wilson, Mike Tresenrider, and Jeff Broughton for reading, editing, and discussing the chapters.

Finally, I would like to thank all the scientists presented in the book. Their efforts in science and their compassion for the sick kept me unflaggingly inspired year after year.

—Billy Woodward

Introduction

ONE DAY in 2002, I awoke feeling like I had a sprained ankle and immediately knew I was in for another bout of gout. Gout intrudes when needle-like crystals form in joints. While painful, it is not life threatening. My doctor prescribed the usual anti-inflammatory drugs, and I waited patiently for the condition to improve in a week, as it always had before. It didn't. My ankle healed, but my hand locked into a crab-shaped claw. The slightest movement—picking up a piece of paper between my finger and thumb—sent pain shooting through my hand. My doctor gave up and sent me to a rheumatologist who put me on steroids. My hand improved, but then my instep became inflamed. I limped along on crutches.

Months went by. Always there was excruciating, throbbing pain in one swollen joint or another. I could not sleep; I could not work; the pain dominated my life. My other joints became inflamed, resulting in constant low-level background pain. Crossing my arms was uncomfortable, for I could not rest one arm upon the other. Each morning, as I gingerly placed my feet on the floor, I wondered—would I be able to walk this day, or even move my computer mouse? During my next trip to the rheumatologist I heard alarming words—"Yours is an atypical presentation." I became depressed, resigned to living the rest of my life with arthritic pain.

After two years of constant pain and continual changes in treatment, one drug—allopurinol—finally succeeded. Within a few months, even the low levels of pain began to dissipate. To sit outside without pain and feel the wind on my skin was extraordinarily sensual. I could take walks again; I could pet a dog. For months I reveled in my newfound freedom. To be pain-free was so exhilarating, to be alive was once again such a joy, that I came to call my cure a miracle. I was living a miracle.

Only slowly did I become aware of how oblivious was my bliss. Someone had saved me from a life of constant pain and I hadn't even bothered to learn their name. The drug I toasted each morning, my elixir, did not spring from a pharmacy. Someone had created it. But who? My shame prodded me to begin studying the process of drug creation, and I learned that drugs are created through arduous experimentation and investigation, over many years, often decades.

Trudy Elion was a woman who fought through much adversity. She never married—her beloved fiancé died from an infection only a few years before the development of penicillin, which could have saved him. Elion devoted her life to science. Coming of age in the 1930s, women were not supposed to obtain doctorates or go into scientific research. Only the shortage of men resulting from World War II swung open the laboratory door for her. Once there, Elion's scientific production was prodigious. Working with George Hitchings, she discovered numerous drugs, two of which save lives to this day—one for childhood leukemia and one that prevents the rejection of kidney transplants. Hitchings and Elion even won the Nobel Prize. They became my heroes, for it is to them that I owe tremendous thanks for developing allopurinol.

As I began thinking about heroes, I wondered how society should choose those it honors with greatness. Why not choose them scientifically, based on evidence? The biggest impact any human can have on another is to save their life. Everything that saved person produces afterward—from progeny, to material goods, to kindness to other human beings—is contingent on their life having been saved. Lives saved can also be measured with scientifically objective criteria. What better metric could there be to determine the greatest people of all time than by counting the numbers of lives they have saved? This book is the result of my attempt to answer a question that, oddly, has never been answered.

Who Has Saved the Most Lives in History?

Such an extensive survey may have been impossible prior to the advent of the Internet, but with access to reams of historical data my eyes were opened to the many brutal causes of death and the remarkable ways humans have found to prevent or postpone it. Surprisingly, all the top lifesavers turned out to be scientists. Politicians and military generals don't top the list because, starkly put, humans cannot kill other humans as fast as disease can. A single disease, smallpox, killed almost twice as many people in the twentieth century as all the wars and genocides combined. The impact of science on the history of humankind is almost impossible to overstate. In the past 500 years, science has had a greater influence on civilization than any other force. Science enables our species to discover the truth about the natural world in which we reside, and thus to manipulate nature to our benefit.

Unfortunately, records kept before 1900 were too meager to measure the lives saved by such luminaries as Edward Jenner, who discovered the first vaccine, or Joseph Lister, who promoted sterile surgery with antiseptics. Even more distressing, I had to leave out Louis Pasteur, who proved that invisible organisms all around us—germs—are the cause of many diseases. As hard as it is to believe now, as recently as 150 years ago much of the educated populace believed that disease was spontaneously generated, and that infections could suddenly arise from the decay of living substances or from nonliving entities like bad air.

Even after focusing the question on the twentieth century, a neatly ordered top-ten list still proved impossible. In 1908, Jersey City, New Jersey, became the first municipality in the United States to chlorinate its water (chlorination had begun a little earlier in Europe). By 1918, more than 1,000 U.S. cities were doing so. Yet the historical sources are vague about how chlorination began and who deserves credit for its use. It is my hope that the website we have set up, scienceheroes.com, will grow to comprehensively cover such less well-documented lifesavers.

A representative list was made, at which time I excitedly hired a biostatistician to tabulate the numbers of lives saved. The concept is simple. If a million people per year are dying from a particular cause, and a cure is discovered that cuts this cause of mortality in half, a half million people per year are being saved who otherwise would have died. Not all discoveries lend themselves to such a straightforward algorithm, so the biostatistician at times applied statistical ingenuity.

When Dr. Amy Pearce, who teaches statistics at Arkansas State University, began turning in the numbers, I was astounded—the lives saved added up astonishingly fast. I had expected the numbers to be in the millions, but when I began to see numbers in the tens of millions, and some adding up to the population of whole countries—hundreds of millions—the importance of these discoveries became palpable. A sizeable portion of the world's population would not be alive today without the ten scientific discoveries cited in this book—and by no means are all of the lifesaving discoveries covered.

In one respect, the vast numbers make sense. The life expectancy in 1900 in the United States was 45 years of age; by the year 2000 it was 77. An immense number of lives had to have been saved to account for such a colossal increase. One of the reasons I hadn't recognized the implication of this monumental advance in life span was that I had assumed that the medicines, vaccines, and nutrition available to me had also been available to my parents and grandparents.

This wasn't true. In fact, modern medicine is remarkably new. The reason doctors made house calls until the 1940s is that they couldn't fix much and could fit most of their medical tools into a little black bag. In practice, doctors were more like hospice workers than healers. Outside of setting broken bones, most of what they did was provide comfort in the form of painkillers until nature took its course, for better or worse. Baby boomers were the first generation to grow up with modern medicine that could actually cure disease.

Who Are These People?

A major surprise was that I had never heard of any of the scientists who saved the most lives. And five of the ten discussed in this book are still alive! Like many people, I consider myself intelligent and well-read. I have browsed *Time Magazine*'s list of the 100 most important people of the twentieth century and various lists of the top 100 scientists who have shaped world history. Not a single one of the scientists who have saved the most lives in the twentieth century are on these well-publicized lists.

Was I wrong about these scientists? Six of the scientists have won the Nobel Prize and all ten have won important science prizes, so undoubtedly their peers have recognized their achievements. Why then are they unknown? Two reasons became apparent. First, for many of these scientists there have been no accounts written about them in the popular press that describe their work in detail. Generalizations wash out in our society—they in no way portray the brilliant insights, the monumental labor, or the magnificent impact of their discoveries. Second, in most of their fields, a mortal problem solved leaves many more unsolved. Most scientists go on working, rather than writing a book or giving interviews. I decided to present these scientists in reverse chronological order and was determined to present their fascinating scientific discoveries in substantial detail.

The book's title was not chosen to detract in any way from the accomplishments of Einstein, of whom I am a big fan. Rather, I chose the title in order to lift these unknown heroes onto the gold medal stand of great scientists, along with their more publicized physicist peers. These scientists had blazing insights of genius, but that was only the beginning. They then put in prodigious hours of work, repeatedly running experiments until they had ferreted out not only the truth of their insights, but the intricacies of the immensely complex human body. Thomas Edison famously said that "Invention is 1 percent inspiration and 99 percent perspiration." What an understatement for these scientists! Making lifesaving discoveries is 1 percent hypothesizing and 99 percent experimenting, then another 1 percent revising prior hypotheses, followed by another 99 percent experimenting, iterated year after year until the germ of an idea bears fruit for all of humanity.

A Culture Can Be Judged by Its Heroes

The thirteenth-century German theologian Meister Eckhart said, "If the only prayer you said in your whole life was 'thank you,' that would suffice." Gratitude is important not only because it acknowledges those who contribute to our well-being, but because it can influence how we invest our resources. Gratitude toward these scientists might inspire children to grow up to be scientists, or it might inspire us to encourage more scientific research.

Gratitude can also improve relationships. Think of taking gratitude to an international level. What if every nation, instead of bragging to its peers about how many nuclear weapons it has, paraded before the world the number of lives it has saved of other nations' people? How different a world would that be?

A culture can be judged by its heroes. By looking at those whom a culture chooses to celebrate, one can tell what its people value. Our modern American culture lavishes attention and wealth on athletes, artists, entertainers, and politicians. These people play an important role in society, so I don't suggest we diminish them. However, let's also celebrate those who have contributed more to society than anyone—the scientists who have saved the most lives in history.

It is a delight to introduce you to the ten hardest-working, most intelligent, most productive people of the twentieth century.

—Billy Woodward

Chapter 1: Al Sommer—
Over 6 Million Lives Saved

The Eye Doctor Who
Discovered a Better Use for Vitamin A

ALFRED SOMMER was hanging onto his seat as he rode the harrowing roads into the highlands of Indonesia. His family was wide-eyed at the new sights. The barely two-lane road was crooked and cramped, much too narrow for the comfort of a doctor who knew the carnage that can litter rural roads. They rose in altitude, with a 2,000-foot drop on one side and a blind switchback up ahead. A car raced past them with nowhere to escape if another vehicle appeared ahead—Sommer grimaced and shut his eyes, but there was no crash of metal, just the laboring of the engine as it chugged on up the mountain. The vegetation had been incredibly profuse in the rainforest of the lowlands—bamboo and orchids and palm trees—and it was green and lush here, too. The road leveled out for a while, then climbed some more, then flattened as they rode past rice paddies, past tea plantations, past tall cinnamon trees. It grew cooler as they rose, slowly heading southeast, toward Bandung, the Flower City.

It was 1976 and Sommer was an ophthalmologist—an eye doctor—who had come to Indonesia to spend three years investigating nutritional blindness. His previous adventures in Bangladesh had been exhilarating, but were muted by the silence of death. He had been working as a medical doctor with a lot of children—cholera and smallpox patients—looking into their big, innocent eyes that implored him to give them life, to heal. It had been rewarding and significant work, but he had not been able to heal them all and there had been too much death. Sommer looked over at his children and a smile creased his face. This time he would be looking into the eyes of children suffering from nutritional blindness, a disease with no fatality rate. He would fight that blindness as hard as medical science would let him, but there was little to suggest it was something they would die from.

Beer Soup

For a mother, the first symptom of nutritional blindness in her child is often night blindness. The inability to see well in the dark was described in ancient times, as was the cure. Hippocrates recommended eating animal liver, which actually helps, although we don't know how he or others learned this.

By 1816, scientists knew there was a direct relationship between blindness and nutrition. Scientists reported that dogs starved on a diet of sugar and distilled water suffered ulcerations in their eyes. During the American Civil War, when a proper diet was sometimes hard to obtain, some soldiers were thought to be malingering, refusing to go on night patrols. They were actually night blind. It was found that their pupils were unable to constrict normally at night when examined by candlelight.

Other eye disorders seemed to be associated with nutrition. A condition referred to as "dry eyes" was described in Europe and seemed to be prevalent in Russia during the Lenten fasts and rampant during the Irish potato famine. Dry eyes appeared to be a precursor to blindness. The cure, discovered in 1881, became the bane of children everywhere: cod liver oil. Word got out, and mothers across America inflicted vile-tasting cod liver oil on their children daily, assuring them it kept them healthy.

As it turned out, Mother actually knew best. There was something in this fatty liver oil—other than its ability to provide energy—that was necessary for health. In 1913, Elmer McCollum and Marguerite Davis at the University of Wisconsin isolated the substance, still without a formal name, and in a famous 1917 study, Denmark's Carl E. Bloch determined much of the substance's health benefits.

Bloch, a pediatrician at the University of Copenhagen, was in charge of an orphanage housing eighty-six children, all under the age of two. Each of the orphanage's two buildings housed two separate wards. One ward in each housed newborns, the weak, and the ill, and the other was home to healthy children. Children in the healthy wards were supposed to be fed the same diet, which included gruel, cocoa, a traditional Danish beer-and-bread soup, and boiled fish or minced meat. Only vegetable oils were used. The beer soup was mixed with whole milk—an altogether healthy diet.

While checking the health of his charges one month, Bloch noticed something wrong. Half the healthy children had night blindness, as evidenced by having dry eyes, and every one of them was in one building. There were no cases in the second building, or in the two wards of sick children. Dry eyes were not rare in Denmark then, during the dark days of World War I, largely because a German submarine blockade hampered food delivery. What was different? He found that the healthy wards' diets differed in only one way: breakfast. The matron in the first building, reacting to an intestinal disorder in some of her children, substituted oatmeal gruel with rusks (a biscuit) for the beer soup. As this had been the diet for several months, the healthy children in that building had received no whole milk for a prolonged time.

Bloch ordered the matrons to continue their diets exactly as they had, other than to feed the children with dry eyes ten grams of cod liver oil twice daily. Within days the children's eyes cleared up.

Researching the problem, Bloch zeroed in on fat in the diet and concluded that the key factor was the "fat-soluble accessory factor A" in the milk, soon to have its name simplified to "vitamin A." As a result of Bloch's work, the Danish government rationed butterfat, guaranteeing access at affordable prices to all Danish children. The problem disappeared in Denmark. Bloch's published research had a brief impact, then was largely forgotten. The developed countries simply improved the diet of their children and night blindness went away. In the less developed world it became accepted by the medical community that eye disorders were the earliest manifestation of vitamin A deficiency.

Following in Forrest Gump's Path

Forrest Gump, of movie fame, had a propensity for being present when historically significant events occurred, such as Elvis learning to dance, John Lennon writing the song "Imagine," and the Watergate break-in. Alfred Sommer never met Elvis and his IQ is on the opposite end of the spectrum from Gump's, but he has happened to be on location for three of the biggest medical advances of the twentieth century.

Meeting Dr. Sommer leaves quite an impression. It isn't his appearance that is so remarkable—he wears professorial glasses and has a recessed hairline—rather it is his personality that is memorable. Known as Al to his friends, when he meets someone, he flashes an easy smile and starts talking. And, boy, does he have stories to tell. To Al Sommer, medicine is an adventure, not a job. He has traveled around the world for the United States Centers for Disease Control (CDC), the World Health Organization (WHO), and Johns Hopkins' illustrious school of public health (now the Johns Hopkins Bloomberg School of Public Health), often living in rough, sometimes even dangerous conditions, and usually taking his family along. When something new interests him, he seems to just decide it's time for another adventure, and off he goes.

Sommer began his life in Brooklyn in 1942, and moved with his family to Queens as a teenager. He had a great interest in history, so he didn't bother with New York City's famed science high schools. He liked to draw maps, and sometimes as he sat drawing them he could picture himself as a professor in a little New England college town. The sixties were just starting when he headed off to Union College, a small, then all-male college in upstate New York. He chose it because a friend was going there and he knew if he went to a small boys' school "it would keep my raging hormones in control and I could get some work done."

Graduation for Sommer was just a way station, for he had always known where his future lay. "My becoming a doctor in the first place was very easy," he says. "I truly believe that my grandmother, on the day of my birth, imprinted it somehow on my soul and, in fact, I can never remember a time when I did not want to be a doctor. She was somebody who had been an immigrant to the United States from Eastern Europe and thought that physicians were people who did good, and did well while doing good, and thought that her grandson couldn't choose a better career for his life. And it was never a tussle. Nobody ever repeated it. I was never threatened or bludgeoned into doing this, but as I grew up and met physicians, as we all encounter our pediatricians and our internists, I couldn't think of anybody who was having more fun because they were, in fact, respected, they were thoughtful, they were intelligent, they were knowledgeable, and they were helping people."

Sommer married his high school sweetheart, Jill, three weeks before entering Harvard Medical School. They shared the optimism of the early 1960s: "I received my MD training at the time that Kennedy had been president. We were all, of course, traumatized and we can all remember where we were the day that he was killed, but we had been inspired. I remember his inaugural address, as does my wife: 'Ask not what your country can do for you, but what you can do for your country.' That was inspiring to young people in those days."

Sommer and his wife both wanted to join the Peace Corps, but there was a problem—a change in the draft laws that would have pulled him out of the Peace Corps. His friends and colleagues suggested that instead he apply to work with the Centers for Disease Control and Prevention, an alternative form of service. He was accepted into the Epidemic Intelligence Service and was soon off to the CDC's Atlanta headquarters for the one-month introductory course. He was assigned to an office for vaccine trials, then engaged with the new rubella (measles) vaccine.

Sommer was, he said later, unimpressed with many of his colleagues, who, it seemed, were far too interested "in the price of eggs in the commissary." He started firing off proposals for initiatives he thought the office should launch. The torrent of requests made everyone nervous.

"Al," his boss said, "we have all voted for you to go home for two weeks and relax." He went home to his pregnant wife and made baby furniture. After he returned, a group of CDC employees who were working in East Pakistan on cholera research came through looking for an additional person to join the group. Except for two things, East Pakistan sounded to Sommer like a good adventure: "The two things I really don't like in life are hot, humid conditions and rice. And then I looked up in the back of the *National Geographic Atlas*—they always list the humidity and the temperature—and it was 120 degrees, it seemed to me on average, in the shade, and 100 percent humidity and I said, hmm…." But he realized the people "were excited about what they were doing"—they were his kind of doctors—so he signed on.

Introduction to Epidemiology

Al and Jill Sommer went off to Dacca, East Pakistan (soon to be re-named Bangladesh) with their five-month-old baby, and Al soon confirmed the truth of the *National Geographic Atlas*: "Well, you learn to like rice and you learn to live in hot, humid conditions. But you come into contact with marvelous people—marvelous people locally who are very much committed to moving the country forward and marvelous people who have gathered from around the world in order to work in places like this. It was a rather extraordinary social and professional experience, and that changed my whole orientation towards medicine and towards the areas of research that I wanted to carry out."

Sommer began by doing basic cholera research, which included answering the question of whether vaccinating family members of cholera victims helps in preventing them from getting the disease (the answer: no). He also examined different types of cholera to see if safe water supplies protected people from each type equally: No, safe water protected people against some types more than others. Another question was the effectiveness of cholera vaccination during an epidemic; the answer: not very effective.

With W.H. Mosley as lead author, the team summarized their research in an academic article: "With the present state of knowledge, the most effective measure for cholera control is the establishment of treatment centers as a lifesaving measure. These can be as economical as immunization programs, and almost 100 percent effective." (They were referring to a new treatment discovered in Dacca called oral rehydration therapy.)

On November 8, 1970, a cyclone darkened the sky over the Bay of Bengal. Later it moved over the mouth of the Ganges River, looming over a huge swath of the delta around the city of Bhola, about seventy miles south of Sommer's home. The storm surge, which coincided with high tide and combined with strong winds, devastated the country. The Cholera Research Laboratory helped out any way it could. After initial aid had been dispensed, Mosley, the center's director of epidemiology, assigned Sommer to do a study on the impact of the disaster.

Sommer coordinated ten two-man teams that went door-to-door, visiting 2,973 families beginning two months after the storm, recording mortality, injury, shelter, and food conditions. They found a staggering 240,000 people had been killed by the storm, including more than 29 percent of all the children under age four and 20 percent of all seniors in the affected area. 180,000 homes had been destroyed and 600,000 people lacked adequate shelter. After two months, one million people were still dependent on outside food relief for survival. It was the biggest epidemiologically documented natural disaster in history.

Sommer loved the work. He says it was "an area of medical research and investigation and endeavor that I didn't know very much about and literally fell in love with—epidemiology—which, in its best sense, is medical detective work. It's Sherlock Holmes played out in the medical arena and clearly has had, and still has the opportunity to have, a positive impact on the lives of literally millions of people at any one time."

Although Sommer was far away from the Vietnam War, another war loomed nearby. In 1971, perhaps partly due to the government's response to the Bhola cyclone, a civil war broke out in East Pakistan—the Bangladesh Liberation War. Before being evacuated with all the other Westerners, Sommer made multiple trips to the countryside, hiding Bengali intellectuals in the trunk of his car, and letting them out where it was safe to cross the Indian border. He often encountered roadblocks along the way, manned by tough, AK-47-toting militants from the country's Northwest Frontier province, forcing him to deploy his smile and ease of talking to spirit them through.

Compulsively Driven to Collect Data

The war lasted nine months, resulting in the new, independent nation of Bangladesh, which was almost immediately confronted with a smallpox outbreak. The World Health Organization requested assistance from the CDC and Sommer joined the hastily assembled team. He was told to organize containment activities, which meant vaccinating people around local outbreaks to isolate the virus.

Sommer remembers it as: "A horrific time. The country had just concluded a civil war (East Pakistan vs. West Pakistan). Ten million Bengali refugees were streaming back home, and they were displacing a large number of Biharis, who had sided with the West Pakistan army. The latter were now gathering in refugee camps. I was dealing with smallpox epidemics in both urban neighborhoods and dense refugee camps. The only thing you could do was pray for early detection and then mass immunization to prevent the disease.

"One-third of those who contracted the disease would die. I mostly worked in a southern district that took a good twenty-four hours to reach, except when I got the unusual airlift out by a Russian helicopter involved in relief work. The key to success was immunizing everyone in the camps. Many refused immunization; I had to resort to holding back ration cards until they agreed to vaccination."

Sommer was ever-observant while the epidemic festered: "Since I'm compulsively driven to collect data without thinking about it, I had these teams not only vaccinate but collect data. I was able to demonstrate that if you got vaccinated within five days of exposure you were completely protected from smallpox, and if you are inoculated within eight days, you wouldn't die of smallpox." This information was later useful when the United States was debating vaccination of hospital workers for smallpox after the 9/11 terrorist attacks. Sommer's evidence that post-exposure vaccination was effective allayed government fears, and as a result, large-scale vaccination did not take place. Along with thousands of healthcare workers, Sommer helped to wipe smallpox from the face of the earth.

During this time, Sommer learned of a thwarted study of nearby children. Nutritionists had long sought a way to measure malnutrition in children. Some obviously effective methods, like measuring body fat or protein content of blood, were too technically difficult for widespread application in the developing world. The most widely accepted method at the time was comparing a child's weight to his age, but exact age was often unknown and frequently impossible to estimate. Weight-to-height ratio would also work, but how do you get scales to rural areas in the developing world? In the 1960s, Quaker medical missionaries in Nigeria had developed a physical test by which they measured the circumference of a child's upper arm (a good indirect measure of muscle mass) against a height table to determine who might be undernourished. Nothing had to be recorded—they could put the height table on a stick so that analysis could be instantaneous.

Health workers in East Pakistan a decade later had accumulated arm and height measurements for 8,292 Bengali children as part of a study, but there was no follow-up. In his spare time, Sommer wondered what had happened to those children. He tracked down as many as he could. Combing through death records, he found that 2.3 percent had died, and in a taste of what would follow years later, he made a startling discovery: During the first month after the children had their arms measured, 1- to 4-year olds in the lowest nutritional category (based on the ratio of arm circumference to height), were at almost 20 times greater risk of dying than their peers whose ratio was considered normal. Predicting the mortality of children could be as simple as measuring their arms! Sommer had trouble convincing anyone to believe his finding and even had a difficult time getting his study published. It was too simple and his data set was too good to be believed. "They never thought of clinical issues like death," he said of doubting nutritionists. It took a while before anyone bothered to replicate the work. Today, arm circumference is a standard measurement.

Bangladesh has a way of changing people. Sommer says of the experience: "It was fantastic, even though we had a cyclone disaster, a civil war, a smallpox epidemic, one thing after the other. It was fascinating. Another culture. Other people. To be fascinated with what you're doing—I went overseas and had this marvelous experience working in the field and literally having millions of people's welfare and lives hinging upon decisions that I made and investigations that I carried out. I found that extraordinarily exhilarating—that the amount of leverage, the amount of impact I could have was so much greater than I could on a one-to-one patient-physician basis. As fulfilling as that is in its own right, I then made the commitment to public health. Public health research is an area where I knew that I was going to enjoy myself."

Chance Favors the Prepared Mind

Some people go down well-planned paths and can see their life unfolding far into the future. Others, like Sommer, take a different approach. "I ought to provide a disclaimer," Sommer says, "some people are far more directed and deliberate than I am; I have lived most of my career by doing whatever seemed most interesting at that particular moment. So I have not said, gee I ought to do this because that'll best position me for the future."

He had intended, on returning to the United States, to become an eye doctor because it was a wide-open field with very few full-time academicians. "I was supposed to begin my training in ophthalmology, but decided that since I had been doing epidemiology for three years, I really ought to learn something about it. And so I again postponed my residency training for an additional year in order to pursue a master's degree in epidemiology at the Johns Hopkins School of Hygiene and Public Health. And that, indeed, made a large difference because I came to ophthalmology with a whole different perspective in which I would think about populations of people and populations of eyes—much more so than the variation you see with one individual."

While there, Sommer's favorite saying of Louis Pasteur's—"Chance favors the prepared mind"—played a hand. A professor hooked him up with Susan Pettis, "a remarkable woman." Pettis didn't get her doctoral degree until she was in her fifties, after which she became the director of prevention for the American Foundation for Overseas Blind (now Helen Keller International). The organization was looking to widen its research and an energetic young "whippersnapper" fit the bill. Pettis' ideas interested Sommer and soon he was hooked. While he was completing his residency, he began helping to design programs and evaluate issues. Sommer set up small studies in Haiti and El Salvador, and he began learning about night blindness.

Al Sommer

What's So Important About Vitamin A Anyway?

The short answer—everything. As Sommer would later describe it, "A child who is night-blind in a village in India or Bangladesh or Nepal literally can't fend for him or herself. While other kids are walking around the village or playing with toys, these children huddle in a corner." Infants rarely show signs of night blindness, but in older children it is obvious to all—their behavior changes drastically at dusk. And the consequences of leaving the condition untreated could be tragic: "The children will go truly blind, because what happens is the cornea, that clear front of the eye, just melts away. And it can melt away in the course of one day."

Night blindness is one of the first symptoms of a condition formally known as xerophthalmia, which is Greek for "dry eyes." (The term is often simplified to "nutritional blindness.") The eyes dry out. Then, waxy spots known as Bitot's spots appear on the conjunctiva, the membranes that cover the whites of the eyes. The dryness spreads over the cornea, causing it to ulcerate and shrivel. The victims are consequently blind for life.

At the time Sommer began studying the problem, as many as half a million kids a year were losing their sight. It was well known that vitamin A deficiency caused nutritional blindness, so some scientists had recommended biannual vitamin A supplementation as a preventative measure—but this had never been proven in controlled trials. When a child with symptoms was presented to the medical community, a shot in the arm of vitamin A was the accepted treatment, and it worked very well if administered before the cornea was permanently damaged. In fact, nutritional blindness responds to a dose of vitamin A within 24 hours. Unfortunately, this treatment was rare.

Vitamin A is not one substance, but a generic term for a variety of related compounds. Briefly, they include the active forms retinol and retinal, and retinoic acid, which can be produced from retinal. Vitamin A is found in animal products, including meat, fish oil, and milk.

Vegetables can also provide vitamin A, but only through precursors such as carotenoids, which are named after their most famous source, the carrot, and include the well-known molecule beta-carotene. They are broken down to form retinal in the intestine. The ability of the body to absorb carotenoids from food varies widely depending upon the way the food is prepared and consumed. Because carotenoids are fat-soluble and often strongly bound in raw vegetables, absorption increases significantly when vegetables are cooked in fat. Normally, more than 90 percent of vitamin A is stored in the liver, which is why fish and animal livers, and foods prepared from them, such as cod liver oil, are such good sources of the vitamin.

Vitamin A then, in its different forms, has many functions in the body. Retinal, for example, is the form in which the vitamin plays its well-known role in promoting night vision. Retinoic acid has been shown to play a crucial role during embryological development. In all its active forms, vitamin A acts as a hormone that regulates the expression of over 300 genes, including many that play a role in cellular growth and differentiation. This regulatory role is most evident in its effect on the mucous membranes of a number of organs, including those of the genitourinary tract and respiratory system (there is some indication of malabsorption of vitamin A in cystic fibrosis patients), and the membrane over the cornea of the eye—hence the correlation between vitamin A deficiency and dry eyes.

Immersed in Indonesia

In 1974, the first international conference on nutritional blindness was scheduled for Indonesia. Sommer, a virtual unknown, had no reason to go. But Susan Pettis had different ideas and wrangled him an invitation. Friends who knew Indonesia's culture cautioned him to curb his tendency to talk too much—it was a culture that respected somber dialogue. Perhaps they were improperly characterizing the people, but quieting Sommer was an impossible task anyway. He quickly formed the opinion that it was a "wonderful country" where "the people seem marvelous," and soon he was as effusive as ever, meeting and befriending numerous people, including those at the Ministry of Health. They accepted his loquacious personality and invited him to come to work with them on a serious problem that his background both in epidemiology and ophthalmology seemed perfectly suited for: Due to some unknown reason, nutritional blindness was more prevalent in Indonesia than in most other places in the world.

Back in Baltimore after the conference, Sommer was invited to join the faculty at Johns Hopkins, but he was not the conservative, academic tenure-track type of doctor. When he told them of the opportunity to do research in Indonesia, colleagues warned him that he would lose his place on the faculty and be forgotten. Sommer told them, "No, I have this wonderful opportunity to go overseas and work with people in another culture. I have something to contribute and it is intrinsically a very interesting series of questions that need to be answered."

And so Sommer headed up that mountain road in the Indonesian spring of 1976. The Nutritional Blindness Prevention Project would largely do research that Sommer planned. He cleverly chose Bandung for a variety of reasons. While it was a modern city with a population of more than a million, it was far away from outside influences. Unlike Jakarta, it was not congested—many of its buildings were only three stories high, architectural remnants from colonial Dutch control a half century earlier. But perhaps just as importantly, it was not hot and humid. A ring of mountains surrounded Bandung, Pandayan Mountain at 8,700 feet being the highest, so the climate was mild. Plus, there were plenty of villages nearby in the highlands where studies could be performed.

Epidemiology, one of the most powerful concepts in medical science, is the study of all aspects of disease in populations, rather than in an individual. It ferrets out the causes of diseases and codifies the best treatments by using one of the basic tools of science—measurement. This concept may seem trivial to a scientist, but is often ignored by nonscientists, as evidenced by whole industries of "alternative" medicine that rely on anecdotal stories or studies that do not measure statistical significance. An epidemiologist uses accurate measurement to discover hidden causes of disease or death by isolating both the many variables in a population, and the various outcomes a disease may present.

Sommer's epidemiological plan was to do three integrated studies over three years to try to determine the risk factors associated with night blindness. He would then have one year to study the data.

The studies however, weren't simply about amassing and tabulating data. There was a personal and cultural aspect to them as well, for it was important to understand the people whom the disease affected. By now, both Sommer and his wife had learned to speak Malay, the unifying language of Indonesia, a country which includes a multitude of islands where more than 300 languages are spoken. They put their knowledge to use listening to mothers whose children had eye disorders.

An early problem had been how to objectively define night blindness. The local people knew night blindness well and called it "buta ayam," which translates as "chicken blindness" because it mimics chickens' inability to see at dusk, the signal that causes them to so ardently roost at night.

Sommer learned that "a mother's history that her child could not see at dusk or dawn was as accurate or more accurate than our objective tests. And the reason we know that is that if we compared our objective tests with what she said, her history had a better correlation with the serum vitamin A levels than our objective test had. So, in this population, we were able to know precisely what a child was getting simply by asking the mother. Total surprise. We had not anticipated this at all."

There were folk remedies worthy of study as well. One widely used in Java on children with either night blindness or Bitot's spots consisted of dropping the juices of lightly roasted lamb's liver into the eyes of affected children. Sommer relates, "We were bemused at the appropriateness of this technique and wondered how it could possibly be effective. We, therefore, attended several treatment sessions, which were conducted exactly as the villagers had described, except for one small addition—rather than discarding the remaining organ, they fed it to the affected child. For some unknown reason this was never considered part of the therapy itself." Sommer and his associates were bemused, but now understood why the folk remedy had persisted through the centuries. Liver, being the organ where vitamin A is stored in a lamb or any other animal, is the best food to eat to obtain vitamin A.

The Studies of Children's Eyes

Sommer's team's first formal study was a countrywide survey of 36,060 preschool aged children from areas of the country that were thought to represent 96 percent of the urban slum and rural population of Indonesia. The parents of any child with symptoms of nutritional blindness were asked dietary questions, as were the parents of 20 percent of all children examined. Children in need of treatment were provided it free of charge. This study would set some baseline statistics to compare the other studies to, but its most important purpose was to quantify the scale of the problem so that government and international aid agencies might be informed of how aggressively the problem needed to be attacked.

The second study was set up in the biggest eye hospital in Bandung—the Cicendo Eye Hospital. The plan was to follow every case of nutritional blindness that presented itself so that the researchers could describe the progression of the blindness as the disorder was treated. Over the course of the three years, about 350 children came in who were going blind, and another thousand came in with precursor symptoms. A pediatrician, an ophthalmologist, and a nutritionist, each of whom gathered data, saw every child. The researchers took photographs of their patients' eyes, examined blood specimens, and then treated the children.

Because they conducted their study in a tightly controlled environment, Sommer and his colleagues could do specific randomized studies on different aspects of the disease. For controls, they would head out to the children's homes and examine other children residing in surrounding houses, finding children who were the same age and sex as each sick child. Some of these children were also found to have nutritional blindness, and so were switched from the control group to the nutritional blindness group and treated.

The studies were dynamic. If the researchers couldn't proceed due to unplanned events, they had to adapt. Sometimes this led to new discoveries.

They ran into one roadblock as soon as they began. The accepted treatment, endorsed at the time by the World Health Organization, was to inject children with shots of vitamin A. Vitamin A is naturally fat-soluble, but animal testing showed that using it in a fat-soluble form wasn't effective. As Sommer says, the vitamin A "sat there as a lump and didn't get out" into the blood. Processing the vitamin to make it water-soluble seemed the answer.

The process resembled that of homogenizing milk—the fat containing the vitamin A is distributed in the liquid instead of sitting like cream on the top. The problem, he immediately discovered, was that there was no water-soluble vitamin A available—no one had bothered to make any. It would take months before the pharmaceutical company Roche could produce some and ship it to Bandung. "What am I going to do in the meantime?" Sommer asked.

Sommer had worked in Bangladesh, where it was shown that dehydration from cholera was fought more successfully by patients drinking fluids (oral rehydration therapy) than by having fluids injected intravenously. So he wondered what would happen if he took the fat-soluble vitamin A and, instead of injecting it, squirted it into the children's mouths? He found that it worked remarkably well.

When the shipment from Roche finally arrived, Sommer was presented with a dilemma—keep using the method that worked, or revert to the book treatment. Sommer is one of the more prominent members of a movement to bring evidence-based science into modern medicine. This movement seeks to base medical conclusions on real, statistically significant, verifiable evidence—not guesses, not anecdotes, not traditional methods. Hard as it is to believe, there is no statistical support for many of the cookbook treatments doctors use or believe in, even to this day.

So Sommer set up a controlled study. He established a trial in which sixty-nine children with corneal symptoms of nutritional blindness were given 200,000 IU of oil-based vitamin A by mouth, and a matched group of forty-five children were given the book dose of 100,000 IU of water-based vitamin A injected by a shot into a muscle. All were given another oral dose the next day. There was no detectable difference in the children's clinical response to the two methods of delivery. This was a monumental change in protocol. A child who comes down with night blindness or other symptoms of nutritional blindness is in a state of emergency. Time matters. But to receive an injection of the vitamin, an Indonesian child had to be transported, sometimes for days, to a healthcare center, and once the child got there, a healthcare worker had to provide the injection using needles and syringes, which introduced an associated risk of hepatitis infection. But anyone could squirt the vitamin into a child's mouth, and it only cost two cents a dose. Thanks to what at first had seemed to be a serious hindrance, Sommer had found a way to bring healing directly to the patient.

The third study would isolate one geographical area and examine as many factors as possible that might cause nutritional blindness. Sommer and his team decided to locate this study in Purwakarta Regency, West Java, a two-hour drive north from Bandung. The area was chosen because there seemed to be a high incidence of nutritional blindness in the region. A census was performed on six villages, involving a total of 6,598 preschool-aged children, who were then categorized by neighborhoods. Teams consisting of eight enumerators, two nurses, an ophthalmologist, a pediatrician, and a nutritionist measured and examined the children. Because of local taboos, many parents refused to allow examination of babies under three months old, so that data set was substandard. Otherwise, parents were very cooperative—only 1 percent refused to participate.

Those with nutritional blindness symptoms were discovered, and controls for each of the sick children were found nearby, matched for age and sex. Blood samples were taken from all children with symptoms, from the chosen controls, and from 5 percent of the entire population of children.

Then the children were divided into three groups. Children with advanced symptoms of nutritional blindness (corneal symptoms) were treated and removed from the study. That left two groups—those with mild symptoms and those with no symptoms. The number of children who had mild symptoms rose dramatically with age from zero percent under the age of 1, to 7 percent of those four years of age.

A major goal of the study was to find out why some children got nutritional blindness and vitamin A deficiency and others didn't. To accomplish this, the study was designed to be longitudinal, meaning that all the children were reexamined every three months for eighteen months. All such large epidemiological studies deal with the vagaries of human beings, so some children would not be at home when the examiners returned, and others developed symptoms of serious nutritional blindness and had to leave the study for treatment. In the end, 3,481 children completed the whole program. The team conducted a total of 20,885 child-interval examinations.

With the three-year study complete and the data piled in boxes, Sommer moved to London for a year and began analyzing the data and writing up the significant conclusions readily apparent. This is what had drawn Sommer into epidemiology: "As you peel back the layers of the onion one at a time, there's always another mystery behind it, another medical detective story." And sometimes the answers didn't jump off the page at him. "Often the answers come at odd times," Sommer says. "You don't get the insights you need—either the answer or how you are going to approach a question—while you are actively thinking about it." Sometimes they came to him while he slept. "I'll wake up at two in the morning, and I'll say, 'Aha, I know how I'll approach that now.' Unfortunately, a lot of times in the morning what I wrote makes no sense."

It was typical epidemiology. For the first time, the disease of nutritional blindness had been thoroughly studied in the field, and a lot of its factors could be explained.

He found that:

- The incidence of nutritional blindness was 2.7 per every 1,000 children in Indonesia per year.
- That meant that 63,000 cases of nutritional blindness occurred per year in Indonesia.
- If the data were extrapolated to the Philippines, India, and Bangladesh, which had similar dietary and related conditions, there could be 500,000 cases of nutritional blindness per year in these four countries.
- Half of untreated cases of nutritional blindness lead to blindness, so potentially 250,000 children a year could go blind in these four countries if the problem was not addressed aggressively.
- Half the population of seemingly normal rural Indonesian children had low levels of vitamin A in their blood—this explained why night blindness was a big problem there.
- Breastfeeding provides an important source of vitamin A—children at age two who were not being breast-fed were at eight times the risk of developing Bitot's spots compared to those who were being breast-fed.
- Children with nutritional blindness ate fewer eggs, carrots, mangos, papaya, dark leafy vegetables, and fish than the control children.

- Vitamin A deficiency was a neighborhood phenomenon, and rarely occurred in isolated cases. Thus treating whole neighborhoods of children at once might be beneficial.
- Children with corneal disease were shorter than children who had Bitot's spots, who were themselves shorter than the healthy controls.

The discovery with the greatest potential impact was that children could take vitamin A by mouth. This meant that treatment was cheap and could be made available even in remote, undeveloped parts of the world.

Living Inside the Data

His research program finished, Sommer moved back to the United States and finally became a full-time professor at Johns Hopkins. He continued his vitamin A deficiency research, but also did research on glaucoma; in addition, he saw patients and performed surgery one day a week.

It was Christmastime, 1982, and things were slow at the Wilmer Eye Clinic at Johns Hopkins, where Sommer was based. No one comes in for cataract surgery over the holidays; it's an elective procedure. It seemed to him to be a good time to take another look at his Indonesian data. Sommer doesn't believe an epidemiologist should simply ask a question and then have a statistician or computer program determine the answer. "I say 'data, talk to me, tell me what you have to say? You have to know your data, you have to smell it, you have to be in it. If you're not living inside the data you are going to miss the most interesting things, because the most interesting things are not going to be the questions you originally proposed; the interesting things are going to be questions you hadn't thought about."

The week after Christmas Sommer was going over the longitudinal study data that examined every child in a village every three months for eighteen months. He was "peeling back the next layer of the onion, as it were. And as we had anticipated, children who had respiratory disease a month ago were more likely to become vitamin A deficient. Children who had diarrhea were more likely to become vitamin A deficient. Children who had a diet that was poor in vitamin A or provitamin A carotenoids were more likely to become deficient."

This was before the desktop computer age, and Sommer was looking through massive, old-fashioned printouts—the kind with alternating light green and white horizontal columns; he was looking at each cell of information in tables. Here is a sample of what Sommer was dealing with:

Example Data Table of Examinations

Child's Age	# of Exams	Children w/ Normal Eyes	%	Children w/ Night Blindness	%	Children w/ Bitot's Spots	%	Children w/ Blindness & Bitot's Spots	%
<1 yr	1,398	1,398	100%	0	0.0%	0	0.0%	0	0.0%
1 yr	3,678	3,658	99%	8	0.2%	11	0.3%	1	0.0%
2 yr	3,705	3,566	96%	61	1.6%	40	1.1%	38	1.0%
3 yr	3,585	3,324	93%	133	3.7%	53	1.5%	75	2.1%
4 yr	3,651	3,359	92%	180	4.9%	65	1.8%	47	1.3%
5 yr	2,758	2,580	94%	86	3.1%	55	2.0%	37	1.3%
6 yr	1,750	1,629	93%	68	3.9%	38	2.2%	15	0.9%
7 yr	360	340	94%	11	3.1%	7	1.9%	2	0.6%
Total	20,885	19,854	95%	547	2.6%	269	1.3%	215	1.0%

There were lots of ways of looking at the data: He could group it by age; he could group it by sex. He was living the data, looking for questions as well as answers. What proportion developed night blindness or the white Bitot's spots in the next round? How many had night blindness in one examination and still had it in subsequent examinations? He began thinking of the children, tracing what happened to each child with their big, dark eyes, seeing their eyes change as the intervals went by. Many eyes cleared up completely, others appeared with Bitot's spots for the first time—sad, innocent, imploring eyes looking for a healer.

As he immersed himself in the information, "It became apparent that something very funny was going on with the data." Children who had night blindness seemed to disappear from the study. They wound up in the "don't know" category far more frequently than did kids who had normal eyes. What happened to those children with their big eyes? Why did they leave the study?

Usually 90 percent or more of the children showed up for the next examination, although occasionally children were in the fields or traveling with relatives. But these missing kids were not kids with healthy eyes. Their parents knew that another free examination was scheduled. Wouldn't they have made sure their child showed up? "Was it because they were too sick to show?" Sommer wondered. "No, their mothers would want to get them help if they were sick. Were the parents of the blind too uncaring, too busy to show?" Not hardly—Indonesians were wonderfully loving parents. Where had the sick children gone?

"Holy cow!" Sommer exclaimed. He suddenly realized the children were not doing the same thing the controls were doing when they missed examinations. The missing children weren't out working in the fields.

Sommer realized the children weren't showing up because they were dead!

"I got really excited and, using my little hand calculator, I redid the data by hand, going from cell to cell. What was the risk of kids with night blindness dying? What was the risk of kids with Bitot's spots dying? How about kids with spots and night blindness? Kids with normal eyes? I checked them through six intervals."

Night blindness wasn't a fatal disease, so the team hadn't been measuring death as a variable. Sommer called in a statistician to rerun the data and make sure his results were correct. They were.

But Sommer knew not to leap to conclusions. Data sets are full of variables, so he began to examine the data to see if causes other than vitamin A deficiency might be the fatal culprit. He ruled out respiratory and diarrheal diseases, which were common among the children. Age wasn't the causative variable either. Nor was wasting due to malnourishment. In fact, malnourished children with adequate vitamin A were less likely to die than well-nourished children who were deficient in vitamin A. The only consistent cause of death, no matter how he grouped the children, was vitamin A deficiency.

Sommer started realigning the data to include the number of deaths.

Mortality of Children in Study

Age	Children w/ Normal Eyes	Deaths	Deaths per 1000	Children w/ Nutritional Blindness	Deaths	Deaths per 1000
<1 yr	1,398	11	9	0	0	na
1 yr	3,658	30	8	20	2	100
2 yr	3,566	36	10	139	9	65
3 yr	3,324	18	5	261	6	23
4 yr	3,359	6	2	292	5	17
5 yr	2,580	2	1	178	0	na
6 yr	1,629	3	2	121	2	17
7 yr	340	1	3	20	0	na
Total	19,854	107	5	1,031	24	23

It was obvious that he had found something new: "When you summed it all up, the kids who had night blindness were dying at three times the rate of kids who had normal eyes. The kids with spots died at seven times the rate, and the kids who had blindness and spots were dying at nine times the rate." Evidently the whole paradigm they and the worldwide medical community had worked under had been wrong. It wasn't at all that changes in the eye were the early symptoms of vitamin A deficiency. Nutritional blindness was, in fact, a sign of advanced deficiency that was causing the children's immune system and organ tissues to deteriorate and thereby allowing them to die of other causes. It was likely that vitamin A deficiency affected health long before it became manifest in the eyes.

Summary table

Exam Results	Number	Deaths	Deaths per 1000
Normal	19,889	108	5
Nightblind	547	8	15
Bitot's Spots	269	6	36
NB & BS	215	10	47
Total	20,920	132	6

The Medical Establishment Turns a Blind Eye

Sommer's finding became the lead article in *Lancet*, the prestigious British medical journal. Then a remarkable thing happened: "Nobody paid any attention to it. Lead article," Sommer says, "no attention. Not one letter to the editor. Nothing."

According to the historian of science Thomas Kuhn, scientists are mostly a conservative lot (as are most of us when it comes to changing our beliefs) who have set ideas about how things work. Kuhn popularized the use of the word "paradigm" to describe this common wisdom, which is what everyone "knows" or assumes to be true. A virus causes smallpox; the sun revolves around the earth; gravity makes apples fall—things like that. Of course, the sun doesn't revolve around the earth, so sometimes the common wisdom is wrong—even if people have accepted it for 500 years. Kuhn calls that realization a "paradigm shift," a rare but crucial concept for the advancement of science. Paradigm shifts come with significant disruption. Doctors are reluctant to embrace them because in medicine the wrong paradigm can kill you, so doctors often find it safest to repeat the past.

The same thing had happened with Sommer's other big discovery—that prescribing oral vitamin A to heal nutritional blindness is as good as an injected shot. The medical community continued to recommend an injection even though a vitamin A pill cost only two cents and, being an oral application, could be easily supplied to a child by any adult. Sommer recalls, "They said, 'Oh no, we can't change the recommendation.' I asked why we can't change the recommendation, and they said, 'People like to get injections.' I said that was a pretty stupid reason." Sommer's finding was still being ignored and would take another five years to become accepted practice.

The nutritionists had their own paradigm, too—that children were protein-energy-malnourished. They even had an acronym to encapsulate all the research they had conducted on it—PEM. It wasn't possible that an outsider to the profession could come in with a magic bullet, a simple vitamin pill that would save children's lives. But the nutritionists hadn't read the history of nutrition, including Carl Bloch's work in Copenhagen with the beer soup. According to Sommer, "A profound amnesia appears to have settled over the broader context of vitamin A deficiency once it ceased to be a major concern of wealthier nations." There were plenty of animal studies and clinical observations that delineated its important role for the immune system and organ tissue growth, but everyone thought that vitamin A only affected the eyes.

Bury Them in Data

Sommer was disappointed in the reception to his discovery, but he isn't the type to sit on his hands. So he did another study, this one in Aceh province, Indonesia, which later would become famous for the 2004 tsunami. Four hundred and fifty villages on the northern tip of the island of Sumatra were randomly assigned to participate in a vitamin A supplementation scheme for one year. Two hundred and twenty-nine villages would have all of their children given 200,000 IU of vitamin A, then six months later the children would get another capsule. Two hundred and twenty-one villages would act as controls and would receive no vitamin pill. A total of 25,939 children were examined at the beginning of the study and again a year later. This time, the findings were even more dramatic. Those receiving the vitamin A pill had a decrease in mortality of 34 percent. (Sommer had thought his earlier study had underreported the benefits of vitamin A supplementation, since, whenever a mild case progressed to a serious case, the child was treated, and thereby some of the children's lives were saved serendipitously.)

Again, *Lancet* published his study, this time with an editorial of support.

"Now everybody noticed," he said, "but they were all angry. There is this thing about when something new comes along, totally unexpected; it's almost the equivalent of getting bad news, like being told you are dying. First, you go into denial, then anger, and eventually you are reconciled. That's what happened; we went through those stages. People should have run out to replicate the findings—do a randomized trial."

One problem was that Sommer's data sets were too good. One critic suggested that if Sommer had only claimed a 10 percent impact, perhaps more people would believe him. Then there was the placebo problem. Sommer had refrained from using placebos at the Indonesian government's request. He says, "I didn't have difficulty with that because placebos, for the most part, counter the potential for subjective bias, where the patient feels better because they've gotten the medicine. But in this instance, death is a pretty hard end point and nobody is going to fake death because they received the placebo. They didn't receive those placebos so, to me, it did not detract from the rigor of the study. But, I must tell you that from a political point of view, many scientists and certainly all policy-makers, who had to make decisions whether they were going to use and divert some of their limited resources to controlling vitamin A deficiency, made a big thing about the lack of a placebo and said they didn't believe the results of this study because it did not have a placebo control."

Others said that vitamin A supplementation might not be safe. But studies showed that side effects are transient and very rare.

There was one party that did not join the chorus of criticism. The Indonesian policymakers had seen enough—they immediately drew up plans for a nationwide vitamin A intervention program.

A wise colleague pointed out that these were the normal reactions to any unexpected research finding. He advised Sommer he needed to "bury them in data." That's what Sommer did. He started another study in the Philippines, but after he had put in years of work and $2 million had been spent, a local guerilla movement forced his team to abandon the area. Sommer moved the operation to Nepal. At the same time, he encouraged other researchers in other countries to try to replicate the findings—sometimes providing direct support.

The study in Nepal showed a 40 percent reduction in mortality. A study in India, where the children were given vitamin A once a week, showed a 50 percent reduction in preschool mortality. Ghana showed 23 percent.

Sommer thought that vitamin A might help fight measles as well as night blindness, since "measles is a viral disease that infects and damages epithelial tissues throughout the body." Vitamin A, in the form of retinoic acid, is known to be important in the maintenance of these tissues, which form the mucous membranes in the intestine, lungs, and eyes that provide the first line of defense against infection. Vitamin A also plays a role in the development of white blood cells, crucial to the maintenance of a strong immune system.

In Tanzania, Sommer found a missionary physician who reported that significant numbers of children with measles were going blind. Their corneal ulcers were easily fixed with the little pills, but it was the perfect opportunity for a study on vitamin A's effect on children with measles. In a randomized trial, half the children brought in with severe measles and without eye lesions were given vitamin A on two successive days, while the other identical half received standard treatment alone. The little pills cut the mortality rate in half.

That paper was published in the *British Medical Journal*, and this time the world was paying attention. A colleague who had earlier debated Sommer publicly, claiming that vitamin A treatment was "too good to be true," soon wrote an editorial in The *New England Journal of Medicine* titled "Vitamin A—Too good not to be true." Within months, the WHO and UNICEF issued a recommendation that all children with measles get vitamin A as part of their treatment. The vitamin works so fast it's effective even if measles is already raging in the child.

Other studies showed that vitamin A deficiency also increased the severity of diarrhea, another major cause of childhood death, and that the little pills reduced mortality from this affliction, as well.

There were still critics, but by this time Sommer believed that any further dickering was unethical. Clearly, the little pills worked and anything that interfered with the treatment was morally unacceptable. He called a meeting at the Rockefeller retreat center in Italy, where he posed three questions. Is vitamin A deficiency bad for kids' health and survival? Is vitamin A deficiency bad for kids with measles, in particular? If you gave kids vitamin A, would you reduce their overall mortality rate and their measles-case mortality rate?

Everyone agreed the answer was "Yes" to all three questions, and the attendees published the results in their favorite medical journals. That turned the tide.

According to Sommer, there are three ways of getting the vitamin into children. The "politically correct" method is to improve their diets, but this is difficult to do. Scientists have recently genetically engineered rice to produce more vitamin A and are attempting to crossbreed it with rice strains native to vitamin A-deficient areas, much as Norman Borlaug once crossed his high-yielding (though not genetically engineered) wheat strains with native germ lines. The idea is that since rice is the staple in much of the world where vitamin A deficiency is rampant, genetically modified rice could mitigate the deficiency problem. This so-called "golden rice" is controversial because it is genetically engineered, and there are doubts as to whether it can provide the amount of vitamin A necessary for children.

Food fortification is another mechanism for introducing vitamin A to the diet, and bread, milk, and margarine have been fortified with vitamin A for decades in Europe and the United States. Countries in Central America fortify sugar with vitamin A. Fortification allows vitamin A to be distributed to virtually 100 percent of the population, regardless of socioeconomic status.

But for much of the developing world, the answer has been supplementation. Vitamin A pills are now standard treatment for children in seventy countries. UNICEF supplies the pills, the Canadian government pays for most of it, and 600 million doses—half of what is needed—are distributed annually. Still, more than 100 million young children don't get the pills and suffer from vitamin A deficiency. But the inexpensive pills make a huge difference for those children who do receive them. According to UNICEF, a half-million children are saved from death every year, which means more than six million lives have been saved since the pills became policy in the 1990s.

The Good Part of Being a Doctor

Sommer has applied his epidemiological approach to other eye disorders. In the 1980s, it was recognized that there were nonscientific methods of diagnosing some eye conditions and determinging varying methods of treatment. Sommer insisted that diagnosis and treatment should be based on evidence, not historically accepted routines. For instance, he took on the problem of magic numbers used to diagnose glaucoma, demonstrating that there was little rationale or predictive power in those being used.

Sommer hasn't stopped with vitamin A research. New research, based on a huge study he and his colleagues have led in Nepal, shows that supplementing pregnant women with vitamin A or beta-carotene, can reduce maternal mortality by 45 percent. It also finds that giving a newborn child a slightly smaller amount of vitamin A within hours of birth dramatically reduces neonatal mortality, when childhood deaths are at their highest.

The story of vitamin A deficiency is a wonderful illustration of the development of scientific understanding over the course of the twentieth century, especially because it brings together findings from seemingly disparate fields of study. It began as a chemistry problem, as Elmer Mc-Collum and Marguerite Davis discovered vitamin A—the hidden ingredient in fat. Then it became a The *New England Journal of Medicine* nutrition problem, as Carl Bloch showed how a diet that lacked it was the cause of night blindness.

Then Alfred Sommer, who obtained his scientist's credentials refining the work on cholera and its solution, oral rehydration therapy, as well as on smallpox eradication, brought to the problem of vitamin A deficiency not only the specialized knowledge of ophthalmology, but also an epidemiological perspective. With it he discovered a whole new paradigm—that vitamin A deficiency is not primarily an eye disorder, but is a life-or-death nutrition disorder. Finally, he realized that being a scientist wasn't enough to begin saving lives. He became an advocate for his own research and for the research of others, until he could convince scientists and policy makers of the life-giving benefit of spending a nickel a year so that each child in the developing world can take two vitamins a year and see into a future.

Lives Saved: Over 6 Million

Discovery: Vitamin A supplements save children's lives.

Crucial Contributors

Alfred Sommer: Recognized that vitamin A deficiency is
                    common in children and supplementation is
                    needed to increase the ability of their immune
                    systems to ward off disease and to help their
                    bodies' tissues be stronger when disease
                    does attack.

After examining
the eyes of
a child with
night blindness,
Al Sommer
prescribed
a large dose
of vitamin A.
Photo courtesy of
Al Sommer.

"The children will go truly blind, because what happens is the cornea, that clear front of the eye, just melts away. And it can melt away in the course of one day."

—Al Sommer

"From existing demographic data, it seems likely that about 125 million children of preschool age are vitamin A-deficient, 1 to 2.5 million of whom die annually."

—Al Sommer

A colleague who had earlier debated Sommer publicly, claiming that vitamin A treatment was "too good to be true," soon wrote an editorial in *The New England Journal of Medicine* titled "Vitamin A—Too good not to be true."

One folk remedy widely used in Java on children with night blindness consisted of dropping the juices of lightly roasted lamb's liver into their eyes.

Al Sommer visits a remote health outpost in Indonesia.
Photo courtesy of Al Sommer.

At the time Sommer began studying the problem, up to a half million kids a year were losing their sight. The team conducted a total of 20,885 child-interval examinations

"I say 'data talk to me, tell me what you have to say'…. You have to know your data, you have to smell it, you have to be in it. If you're not living inside the data you are going to miss the most interesting things, because the most interesting things are not going to be the questions you originally proposed; the interesting things are going to be questions you hadn't thought about."

—Al Sommer

Vitamin A Facts
- Vitamin A is not one substance, but a generic term for a variety of related compounds.
- The ability of the body to absorb vitamin A from food, through carotenoids, varies widely depending upon the way the food is prepared and consumed. Because carotenoids are fat-soluble, and often strongly bound in raw vegetables, bioavailability increases significantly when vegetables are cooked in fat.
- Normally, more than 90% of a person's vitamin A is stored in the liver, which is why animal and fish livers, such as cod liver oil, are such good sources of the vitamin.
- In the American Civil War, soldiers thought to be malingering were often actually nightblind, caused by Vitamin A deficiency.

Al Sommer in the mountains of Indonesia.
Photo courtesy of Al Sommer.

"You come into contact with marvelous people. Marvelous people locally who are very much committed to moving a country forward and marvelous people who have gathered from around the world in order to work in places like this—a rather extraordinary social and professional opportunity on our part and experience, and that changed my whole orientation towards medicine and towards the areas of research that I wanted to carry out."

—Al Sommer

"[It was} an area of medical research and investigation and endeavor that I didn't know very much about and literally fell in love with—epidemiology —which, in its best sense, is medical detective work. It's Sherlock Holmes played out in the medical arena and clearly has had and still has the opportunity to impact positively on the lives of literally millions of people at any one time."

—Al Sommer

In 1971, Sommer was working in Bangladesh when a civil war broke out. Before being evacuated with all the other westerners, Sommer made multiple trips to the countryside, hiding Bengali intellectuals in the trunk of his car, and letting them out where it was safe to cross the Indian border. He often encountered road blocks manned by tough, AK47-toting militants from Pakistan's Northwest Frontier province along the way, forcing him to use his smile and ease of talking to spirit them through.

Chapter 2: Akira Endo—
Over 5 Million Lives Saved

Statins: Life Extension
for the Baby Boomers

EVERY SPRING, the Japanese follow the *sakura zensen*—the cherry blossom front—as it moves northward week by week. It takes months for the cherry trees to bloom over the length of the country, ending in the north where the snow has just melted and the mountain clearings have greened with rejuvenated growth. Akira Endo was born on November 14, 1933, in the Akita Prefecture of northern Japan, and every spring he saw his family's backyard cherry tree bloom. The trees were such an important harbinger of spring that they came to symbolize much of Japanese culture. As World War II ground down the nation, kamikaze pilots painted the flowers on the sides of their planes, believing that as fallen warriors they would be reincarnated in the blossoms. Even today the military and police use cherry blossoms as an insignia the way Americans use stars.

For a boy growing up in rural Japan during World War II, the evidence of war was not so much what came into the region, as what was taken from it. There were no bombs, no invading army, no war factories. But sucked from the region were men, food, and carefree youth. Akira Endo was nine years old when America first bombed Japan, and in his autobiography he remembers the effect the shortage of working men had on children: "When I was in the upper grades of elementary school, we took part in labor service projects called *kinrou houshi* to help the local farmers. At the labor service all the classmates helped the farmers and picked edible wild plants from the fields and mountains near the farms."

Endo was attuned, as farm boys are, to many such natural things on his family's farm, where they grew rice and raised goats, sheep, rabbits, and chickens. Most of his time was spent with his grandparents, who lived with the large family. He slept in their room and listened to their stories. His grandfather was well liked in the community and did fortune telling and amateur doctoring on the side. Since there was no local doctor in the area, he would dispense drugs and cures for minor injuries and sickness. In the autumn, when Endo was in the fourth grade, his beloved grandmother became sick. At night, she took his hand and said, "Grandma's disease is the one which cannot be cured," and made him touch her belly. He would place his fingertips gingerly on her abdomen and feel a small crunchy thing. His grandfather could not cure it, nor could the professional medical doctor they sent for. Day by day the crunchy thing got bigger and Endo's grandmother got thinner.

As with most farm boys, work came first for Endo, before study or play. Amongst his many chores, he had to take care of the chickens and plant a whole field of rice by hand, one root at a time. But in his free time he played in the fields, streams and mountains near the farm. Like all boys, he had dreams of who he would be when he grew up, but circumstances removed lighthearted little-boy dreams from his future. The heroes he wanted to emulate were serious men. One overcame a debilitating burn to his hand to become a famous bacteriologist at the turn of the twentieth century. Endo recalls, "At that time I had several reasons why I wanted to become a doctor or scientist. First of all, I learned about Hideyo Noguchi. He got burned just like I got burned at the age of five, by the fireplace. Secondly, I suffered from pneumonia when I was a 4th grader."

That May, the double cherry blossom tree in Endo's backyard bloomed gloriously, and his grandmother was released from the pain of cancer. He thought, "I want to become a doctor to help people who are suffering from diseases."

It was a solemn and sobering childhood. Endo was eleven years old when the war ended. He recalls, "We lost the war. Even adults did not know what to do. We did not have food and clothes. Everything was called off. Even kids lost their dream. It was so hard that even kids did not have hope."

Japan's total surrender at the end of World War II eroded much of the country's national pride. Arising from the ashes, the young generation would need to build a new Japan. Akira Endo grew up to embrace a modern way of thinking, embodied by science, and with it he crafted

34

a gift to the world that would rekindle Japan's pride in what it could accomplish, and that people from all nations could celebrate.

A Student of Fungi

Bookending the cherry blossoms of spring, the northern Japanese autumn produces a kaleidoscope of color across the mountain vistas. In the forests of falling beech and oak leaves, what most attracted the young Endo lay underneath the trees, beneath the damp, speckled forest floor. "When I was in junior high school, my grandfather often took me to the mountains near my house to pick mushrooms in the autumn. He taught me how to tell the difference between mushrooms and toadstools [in some cultures poisonous mushrooms are called toadstools]. Almost all the poisonous ingredients of toadstools are water soluble, which means that if you boil them, the poisonous ingredients dissipate into the water. So you can actually eat most poisonous toadstools if you really wash them after you boil them and get rid of the broth." Mushrooms were eaten at almost every meal in the Endo household.

Endo made good grades in elementary school, but his morality grade was consistently low. "I was a careless person and not calm," Endo admits of his first years in school. Only in junior high school did he begin to take school seriously, but matriculating on to high school presented a problem. The nearest school was not within walking distance and his family could not afford the cost of boarding him away from home.

Tadao Hatakeyama, Endo's childhood best friend, recalls that one day Endo arrived late to school. When he asked him why he was late, Endo told him, "I negotiated with my father personally about going to a high school. And I rampaged until he said yes." Hatakeyama asked, "What did you do as rampage?" Endo replied with strong, clear eyes, "I wore boots and walked and rampaged on a rolling board [for Japanese rice cake]." Relatives remembered the young Endo as headstrong and weren't surprised he won out.

As a result of his rampage, Endo began attending a work-study school which met only on weekends. It was a two hour walk each way and was not advanced enough to prepare him for college. Fortunately, his uncle moved to the city of Akita and accepted Endo into his household, where he could attend a real high school.

There Endo learned more about fungi, including the microscopic molds that he found on mandarin oranges and koji (steamed rice that is cultured with mold spores so that it can be used to make sake, a sweet wine). He also retained his interest in the larger fungi—the poisonous mushrooms his grandfather had taught him about. One type killed both flies and humans. Another, called the fly-catching mushroom, had evolved a subtlety that Endo found fascinating. Lemon yellow in color, people cut it up and placed it on plates throughout their homes. Flies were attracted to it and later died, yet it was not poisonous to humans.

It was so mysterious to Endo that during summer break he devised his own experiment. He boiled a fly-catching mushroom and put the cooked mushroom on one plate and its broth on another. Flies were attracted to both, but only those that drank from the broth died. "I confirmed that the poisonous ingredient of Tricholoma muscarium Kawamura [which kills flies, not humans] is water-soluble, like my grandfather said," Endo recalled.

Endo loved chemistry and his principal encouraged him to take the entrance exams to Tohoku University. His parents, knowing that if he passed the tests they could not afford to send him to college, sent his older brother to talk Akira out of dreaming so big. Their scheme was based on his clothing. Endo's dress could not be called stylish since he had only one pair of socks and had to wash them every night and let them dry by the fire. Tomiko, his brother's wife, recalls the trip: "His parents asked us to tell him that it was 'not only you; you have brothers and sisters, so you cannot go to a university.' They gave us 20,000 yen [about $55] and said to us, 'Tell Akira if he does not go to the university, we will buy him a suit. Trick him.'"

When they arrived and attempted the bribe, Akira retorted, "I don't need any clothes, let me go to the university. Cut me off from your money." "Cut me off" was a serious phrase meaning "disown me," so they knew Akira had made up his mind. Fortunately it never came to that; Endo's principal helped him get a scholarship.

Steaks the Size of Sandals

Endo lived in the college dorm with three other students in a 10-tatami mat room (a mat being approximately 3' x 6'). In a college photo album, Endo's cheeks appear sunken. The cafeteria did not provide enough food for a young adult, and Endo recalled that, "sometimes I was too hungry to concentrate, and I could not hear a lecture. Also, sometimes I felt dizzy and crouched down in a hallway to keep from fainting." The school had a system the students called *zwei*, for second meal. Some students did not eat all of their allotted food, so around 8 PM a drum would signal that leftovers were available. The drum set off a race to the cafeteria for the remaining food, which was given out on a first come-first served basis. Endo studied from 7 PM to 12 PM every night, but always had his chopsticks ready. "I was good at running," Endo recalls. "I ran like women running for a bargain. After all, I was not full even right after supper."

Though the food was lacking, the knowledge Endo gained was ample and nourishing. At first he studied agriculture engineering, until he came across the biographies of Alexander Fleming and Selman Waksman, pioneers in antibiotics. They reminded him of his earlier interest in molds and opened his eyes to bacteria, leading him to change his major to applied microbiology.

After graduating from college, Endo moved to Tokyo and was hired by the Sankyo Corporation, a pharmaceutical company with products ranging from medical to agricultural. It was founded at the turn of the century by Dr. Jyokichi Takamine, who in 1912 donated the 3,000 cherry trees that now so extravagantly mark springtime in Washington, D.C. Within two years, Endo was doing biochemistry research on his "precious" fungi. After going through 250 species, he discovered one that produced pectinase, an enzyme that decomposes pectin. Pectin is a type of fiber produced in fruits such as apples and grapes which causes their juices to be cloudy. Two years later, Endo commercialized a process to make great quantities of pectinase. It was exactly the kind of applied science Sankyo wanted its researchers to do, and pectinase became very profitable to the company as a clarifying agent.

At that time in Japan it was possible to obtain a doctorate by doing at least five years of research at a corporation after college. Endo wrote fifteen research papers on pectinase, passed the qualifying exams and his degree thesis, and obtained his Ph.D. in 1966. His stellar research and acclaim from the commercial success of his pectinase production earned him a reward—the opportunity to study in the United States for two years.

With his pectinase work winding down, Endo examined his life. He had come a long way. He was married now, with a son, and his career looked promising. He could now spare himself time for something more altruistic. Endo thought, "Since I was born as a human in this world, I want to leave my mark before I die. I want to die after I do at least one thing useful for the world." Remembering his grandmother's sickness, he turned to medicine. Sankyo manufactured and marketed several antibiotic drugs, and the company was open to his changing fields. On long walks, he pondered which field of research to study.

Endo says, "At that time, doing research on nucleic acids and proteins was very popular among other researchers, so I avoided researching those topics. I chose the biochemistry of lipids [cholesterol and fatty acids] as my research theme. The reason why I chose this theme was that not so many researchers were researching it. This meant I did not have much competition."

Endo wished to study at Harvard under Konrad Bloch, who had just won a Nobel Prize for his research on cholesterol metabolism. Unfortunately, Bloch's student load was full. Consequently, in the autumn of 1966, the 32-year-old Endo, with his wife Orie and his two-year-old son Tadasu, moved to New York City to attend Albert Einstein College of Medicine.

The day Endo and his family stepped off the plane they were confronted with the proportions of everything American. Everything was bigger. Their living space was three times the size of their apartment in Japan. Postdoctoral pay at the college of medicine was four times Endo's Japanese salary. And Americans—Endo and his wife were greatly amused when they saw how many people looked like big sumo wrestlers. At restaurants, Endo watched with amazement as people ordered steak "the size of a sandal"—enough to feed a family in Japan. Then they would trim off the fat before eating it. "This was culture shock," he said, "something inconceivable in Japan." The Japanese did not eat much red meat—indeed, Endo had never eaten steak—but when they did, they liked their steaks tasty and well-marbled with fat. Kobe beef, an expensive Japanese delicacy, is almost 50 percent fat, the result of the cows being fed a diet of grain and beer.

Although Endo was studying phospholipids, which are the principal component of biological membranes, he also sought out Bloch, who became one of his most important mentors. "I never had the chance to study at Dr. Bloch's laboratory," Endo says, "but he encouraged me and guided me for more than thirty years. He did this from the time I was at Albert Einstein College of Medicine until a few years before he passed away in the mid-1990s. Although he was a great scientist, he was modest. He was good-natured and had a good heart. I was drawn to him and respected him. If I had never known him, I would have never studied about cholesterol this much."

The Culprit Was Cholesterol, and It Was Big News in America

Endo learned that a high blood cholesterol level was thought by some to be a major risk factor for heart disease. Cholesterol is a fatty, wax-like substance found in the cell membranes of all tissues. Its name is derived from the Greek *chole*, meaning bile, and *stereos*, meaning solid. It is essential to animal life, providing structure and fluidity for cell membranes over a range of temperatures. Cholesterol also plays a crucial role in the production of bile, which helps in fat digestion, and of various steroid hormones such as estrogen and testosterone. It plays other roles as well, including metabolism of the fat soluble vitamins A, D, E and K.

Ninety-three percent of all the cholesterol in the body is in cells, most of which can manufacture their own supply. Only about 7 percent is circulating in the blood at any one time. The cholesterol that circulates in the blood comes from the diet or from the liver, which can synthesize enough to supply any needs the body may have. Cholesterol is water-insoluble, and swishes through the arteries, veins, and heart by hitching rides on proteins called lipoproteins to get where it needs to go. There are two kinds of lipoproteins, a "good" one and a "bad" one.

Low-density lipoproteins (LDL) carry cholesterol from the liver to the rest of the body. It is called "bad cholesterol," a misnomer because

LDL serves an absolutely necessary function; only when there is too much of it is it bad. The second type, high-density lipoprotein (HDL) is the "good cholesterol" because it carries excess or unused cholesterol from the arteries, where it might do damage, back to the liver, where it is broken down and sent for excretion.

Total cholesterol count includes both LDL and HDL. If it is less than 190mg/dl, everything is fine. If it is between 190 and 240, it is considered borderline high. If it is above 240, it is considered high and development of atherosclerosis is considered probable.

When spring rains bring down cherry blossoms, they choke gutters and sewer pipes, sticking to the walls and clogging up the runoff system. Cholesterol in the blood behaves similarly. If there are too many LDL particles in the blood, cholesterol can stick to the walls of the coronary arteries, triggering an inflammatory response, causing pimple-like lesions to form on the artery walls, and eventually blocking the free flow of blood—essentially building a dam in the arteries. This buildup is called atherosclerosis. Calcification with plaque and scar tissue occurs, and the arteries become less flexible, more rigid, and resistant to the life-giving pulse of the pumping heart.

This buildup of plaque begins early in life. In fact, during World War II, doctors autopsying young soldiers were amazed to find that many 20-year-olds already had arteries that were thickening from the start of atherosclerosis. If left unchecked, the reduced flow of oxygen-rich blood can eventually lead to pain in the heart called angina. In addition, if the plaque breaks off, it can lead to blood clots. If a blood clot cuts off enough blood, the heart is starved, heart muscle dies, and a heart attack occurs. The same mechanism causes strokes. A clot forms and blocks the flow of blood to the brain, which causes cells to die due to a lack of oxygen.

Endo learned that the culprit was cholesterol, and it was big news in America. He recalls, "In the quiet residential area of the Bronx where I lived, there were many households consisting solely of elderly couples, and the aging of the population in that area was surprisingly advanced. The appearance of ambulances at any time of the day or night was a common occurrence and I often caught sight of ambulances carrying away patients with myocardial infarction. At that time in the United States, heart disease was the leading cause of death, surpassing the number of deaths due to cancer. It accounted for the deaths of between 600,000 to 800,000 people annually. The number of patients with high blood cholesterol levels, which is a condition which leads to coronary artery disease, was said to be over 10 million people. I was also surprised to find that people in the United States were extremely nervous about cholesterol, which at the time had not become an issue that weighed on the minds of people in Japan. On the other hand, cancer and cerebral hemorrhages were the first and second leading causes of death in Japan, and, therefore, people did not pay attention to either heart disease or cholesterol. However, it was my view that heart disease would increase in Japan in the future as the diet became richer, and cholesterol would then become a serious issue."

Inspired by Bloch, Endo began dreaming of discovering a drug to treat high cholesterol.

According to Endos, "Cholesterol is supplied to the body through two routes: either through the intake of food where it is absorbed by the gastrointestinal tract, or through synthesis by the body, mainly by the liver. When the former supply is unable to meet the required amount, the latter compensates. However, when the supply is adequately met by the former, synthesis of cholesterol by the liver is suppressed. In this way, the body has a mechanism to prevent cholesterol from becoming excessive. By the end of the 1960s, it was already known that the synthesis of cholesterol was controlled by HMG-CoA reductase."

Bloch, along with Fyodor Lynen, had unraveled many of the details of how the body synthesizes cholesterol. The process by which the body modifies an initial precursor molecule repeatedly to convert it into cholesterol is quite complex and involves five to ten major steps; these can be broken down into many more minor steps that include more than 30 enzymatic reactions. One of the key steps involves the difficultly named enzyme HMG-CoA reductase (full name: 3-hydroxy-3-methylglutaryl-Coenzyme A reductase; more simply HMGCR). HMGCR catalyzes the reduction of HMG-CoA to mevalonic acid. HMGCR is crucial in the cholesterol synthesizing process, but is also the rate-limiting enzyme that increases when cholesterol decreases, catalyzing the production of more cholesterol, and vice versa.

Scientists later learned that the body has a feedback system to regulate not only how much cholesterol is made—via the HMGCR enzyme—but also how much LDL is in the blood. Some liver cells have LDL receptors on their surface. These are proteins that can grab hold of LDL and remove it from the blood. It turns out that both genetics and diet influence the number of LDL receptors present in the liver. One in a million people are born with two mutant genes that cause them to have very few LDL receptors. People with homozygous FH (*familial hypercholesterolemia*) have livers that remove very little cholesterol from their blood, resulting in cholesterol levels six to ten times higher than normal (1,000 to 2,000mg/dl). These people often begin having heart attacks as children. One in 500 people have one mutant gene that causes heterozygous FH. They have twice the cholesterol levels from birth of most people, and begin having heart attacks in their 30s or 40s.

For most of human history, obtaining enough dietary fat to survive was a problem for the vast majority of people, and high cholesterol was not a significant concern. Only in the last 100 years have countries' entire populations had ample-enough food resources to have too much fat in their diets. Through a mechanism that is not clearly understood, about two-thirds of people in industrialized countries have genes that cause the body to repress the LDL receptor gene when they consume a high-saturated fat diet. With a diet that encourages the synthesis of cholesterol, but with no increase in LDL receptors to remove the excess, cholesterol levels for two-thirds of the population remain at a high enough level to make clogging of the arteries epidemic.

People can't change their genetics, but they can change their diets, which influence cholesterol levels. However, managing cholesterol isn't simply a matter of will-power. Removing cholesterol from the diet often results in a minimal decrease in blood levels of LDL. Removing saturated fats has more of an effect, but many people possess a genetic makeup such that even on a low-fat diet they can't cut their LDL cholesterol by more than 10 percent, which is not always enough to prevent atherosclerosis.

A Brilliant Hypothesis

Endo returned from America excited about doing research on a drug to lower cholesterol, but was immediately stymied. He says, "For one year after my return, I could not do proper research and got bored at work because there were differences in research policy between my direct supervisor and his boss, the laboratory director, and reserve between my direct supervisor and me."

Then Sankyo made the wise business decision of building upon its proprietary drugs. It set up a new laboratory, the Fermentation Research Laboratory, primarily to search for new antibiotics. Endo was assigned to the lab as a group leader and had complete freedom in research.

It doesn't take long when reading Endo's writings or interviews before his style of thinking becomes apparent. He learns as many of the aspects of the basic science as possible, then analyzes comprehensively. His thinking is deliberate and clear, and he is exceedingly logical.

Endo did not know about the LDL receptors that removed cholesterol (the role of LDL receptors was later discovered by Michael S. Brown and Joseph Goldstein, for which they won the Nobel Prize in Medicine in 1985). However, he did know that the drugs used at the time to lower cholesterol did so by increasing the removal of cholesterol from the body or by inhibiting its absorption from food, that neither method was very effective, and that both had severe side effects.

Endo says, "In view of this factor, I believed that rather than an uptake inhibitor to control the absorption of cholesterol supplied to the body through the intake of food, an HMGCR inhibitor that would suppress liver cholesterol synthesis would be much more effective as a cholesterol-lowering agent.

"There were many people who were critical. They would say things like 'an inhibitor of cholesterol synthesis is dangerous.' They said that the inhibitor of cholesterol synthesis might also inhibit synthesis of bile acid, adrenal cortical hormones, sex hormones, etc., and as a result it could be the cause of side effects."

"I could not agree with these thoughts because the cholesterol synthesis inhibitor is not for healthy people. Instead, it is for patients who are suffering from the accumulation of surplus cholesterol, and we administered it to reduce the cholesterol level to a normal level. Just as people could have a serious handicap if you lowered their blood sugar and blood pressure levels too much, people could have a handicap if you lowered the cholesterol level of healthy people below normal levels. However, there should not be any problems if you lower the surplus of cholesterol and keep the normal cholesterol level. An inhibitor of cholesterol synthesis does not mean it stops cholesterol synthesis completely."

Endo wasn't the first to think about inhibiting HMGCR, the crucial enzyme in the cholesterol synthesis pathway. His brilliant insight was developing a hypothesis as to where he might find a substance that would target the enzyme.

Endo knew a lot about microbes. For instance, he was aware that "during some twenty odd years from the discovery of streptomycin until the end of the 1960s, more than 1,000 strains of antibiotic substances were discovered."

He also understood the role evolution plays in the cutthroat world of microbes: "From the example of the antibiotics, I predicted that there were microbes that could make a cholesterol synthesis inhibitor. There was a theory that microbes make antibiotics to kill other microbes that were foreign enemies, or to inhibit their growth. I took to this theory, and I thought it was possible that a microbe existed which could inhibit cholesterol synthesis."

There were sound reasons to postulate that such a microbe might exist. Plants and fungi produce cholesterol, although in much smaller quantities than do animals. Mevalonic acid, which the HMGCR enzyme helps to produce as a step in cholesterol synthesis, is also a precursor to other biological molecules. Was it possible that some fungi had evolved a way to limit cholesterol synthesis to protect themselves against attacking microbes that might require cholesterol or one of its precursor molecules for growth?

How do you set about finding a substance that has what must be a rather rare trait—cholesterol synthesis inhibition? There were literally tens of thousands of candidates, so Endo culled the less promising ones in his mind. He says, "In those days, Actinomycetes were praised as a treasure of the physiologically active substance, and a large number of Actinomycetes were divided from the soil all over the world and used for treasure hunting."

Researchers were gravitating to Actinomycetes, now more specifically known as Actinobacteria, because they had produced many antibiotics. In fact, one of Endo's scientific influences, Selman Waksman, had studied thousands of Actinobacteria species. Endo knew that simply finding an active substance was just the half of it. He needed an active substance that was safe enough for continual ingestion, because he realized that maintaining low cholesterol might require lifelong therapy for many individuals.

There had been safety concerns with some antibiotics made from the Actinobacteria; for example, streptomycin occasionally produced the side effect of hearing loss. Endo thought, "There is no precedent in which actinomycetes are used for production or processing of food. That worried me about Actinomycetes."

In contrast, some molds and mushrooms are edible. While they had yielded considerably fewer antibiotics, the ones they had produced, such as penicillin, were remarkably safe. Endo says, "I chose mold and mushrooms as microbes which had a strong possibility to produce a cholesterol synthesis inhibitor; I did not choose actinomycetes, although it was against the trend of those days."

Ever logical, Endo remained pragmatic. "With that said, there was no guarantee that I would be able to discover an HMGCR inhibitor, so I decided to study a few thousand strains of fungi over a two-year period. It was research that was like a bet, and if I was unable to discover the substance I set out to find, I would then terminate my research."

Treasure Hunting

While Sankyo's Fermentation Research Laboratory director was on Endo's side, he made it clear there were financial constraints. The company assigned Endo a one-story building to use as a lab. It was old, with a cold concrete floor and rusting iron sashes enclosing drafty windows that in stormy weather allowed rain to creep in.

Endo usually rode the train to work, getting off at the Osaki station and walking ten minutes to the lab, deliberating all the way on how to most efficiently attack the problems of the day. He says, "The key was constructing an experimental method to examine more than 100 specimens at once and to reduce consumption of expensive radioactive material as much as possible. I solved these two problems by constructing a method which scaled down an existing method using the liver enzyme of rats from the traditional volume of 5.0 ml to .2 ml, and by improving examination work, which ordinarily took two days to finish, to be able to finish in one day. We needed four people to do this research, but we did not have enough people because there were just two research assistants (highschool graduates) and me at that time."

In April of 1971, the cherry blossoms bloomed and the company hired a freshly minted college chemistry graduate, completing Endo's team. Finally, he could walk to work in anticipation of real progress.

The First Run

There was no shortage of molds to investigate. Sankyo kept around 2,000 molds and mushrooms at the fermentation laboratory, taken from soil, dead leaves, and fruit. The lab continuously collected molds and Endo could buy additional ones from research institutes around the world. Getting the molds would not be the problem. Analyzing them would be.

Step 1: Grow one hundred culture broths.

The first step was to culture a hundred species of fungi in separate flasks for a week. Soon cloudy culture broths were everywhere.

Step 2: Mix and test radioactivity.

The first test was to see if any of the cultures inhibited cholesterol synthesis. The idea was to winnow the hundred down to a few potential targets. The test involved rat liver extracts. Because rat livers use enzymes to convert precursor molecules into intermediate molecules, then into cholesterol, just as human livers do, a broth of rat liver enzymes mixed with a precursor molecule—in this case acetic acid—produces cholesterol. If the precursor molecule is made radioactive, cholesterol production can be measured, and the production of an unadulterated broth can be compared to that of a similar broth to which the fungal specimen has been added. Any significant difference in cholesterol measured can be attributed to the fungus.

Endo was pleased to find quite a few molds that inhibited cholesterol synthesis. His hypothesis was correct: Some molds had evolved the ability to produce a cholesterol synthesis inhibitor. However, this was just the beginning of the search.

Step 3: Discard molds that have logistical problems.

Not all the substances that demonstrated cholesterol inhibition were potential drugs. There are numerous prerequisites for a potentially therapeutic substance to be considered for actual human use. Obviously, it must be safe. It must also be easily and cheaply reproducible. Endo examined the inhibitory broths for their active substances and discarded those that contained large molecules or that were unstable when heated. Those that reproduced poorly were also discarded. He knew none of these would be effective as human drugs.

Step 4: Discard broths that do not show early stage cholesterol synthesis inhibition.

Cholesterol synthesis is a multistage process. The HMGCR enzyme is active in an early stage that turns HMG-CoA into mevalonic acid, which itself is an ingredient in later stages of cholesterol synthesis. The active culture broths were therefore tested for their ability to inhibit lipid synthesis from mevalonic acid. If they did, they were acting on a later stage of the process, so could be discarded.

Step 5: Test for HMGCR enzyme inhibition.

Each potential broth was added to a solution containing the precursor molecules that the HMGCR enzyme acts upon, mixed with rat liver enzymes (which contain the HMGCR enzyme), then heated to 37C (the temperature of the human body—98.6F) for thirty minutes. After the reaction was chemically concluded, it was reheated to 37C for fifteen minutes and then applied to a silica gel plate. Sections of the plate where mevalonolactone, the by-product that the HMGCR enzyme catalyzes, was located were scraped and measured for radio-activity. This measurement could be compared to that of the control solutions that did not contain the broth to see if the broth had inhibited the HMGCR enzyme.

Step 6: Purify the active substance in the broth.

As Endo explains, "Contained in the microbial culture broths were high-molecular substances such as proteins and polysac-charides; low-molecular water-soluble substances; fat-soluble substances (substances not soluble in water but soluble in spe-cific organic solvents); substances that were heat-stable and unstable in heat; substances that were acid- or alkali-stable or unstable; substances with low-toxicity and high-toxicity; and both known and unknown substances. There would be a mini-mum of 100 types of substances in each case, and in higher incidences there would be over 1,000 types of substances."

To isolate and purify the active substances in the molds, Endo used various organic solvents as well as various chromatogra-phies. Organic solvents are stable carbon-based chemicals that when added to mixtures can separate out various substances. Chromatography passes solutions in one state (liquid or gas) through rows of chemical filters of a different state. Substances pass through these media in different amounts of time, allow-ing them to be isolated from each other.

Sometimes this took a long time, Endo says: "To assess wheth-er an active substance present in a culture broth where inhibi-tory activity was observed could potentially suit our objectives, it was necessary to extract that substance in a pure or refined state from the broth. There were some simple ones where the refining of an active substance could be completed in one week; but there were also some difficult samples requiring a much longer period of up to a year."

Step 7: Discard known substances.

Some of the active substances Endo and his team isolated could be identified as known substances. They usually discarded these since their properties were generally known and could be looked up in books. They focused their investigations on novel substances.

Step 8: Measure the inhibiting potency of the novel substances.

Endo and his team selected for further research only novel substances that greatly inhibited the HMGCR enzyme. They measured potency by the inhibitory effect per unit of weight.

Step 9: Test the substances for toxicity.

No matter how much a substance inhibited the HMGCR enzyme, if it was unsafe, it could not be a drug. They tested all remaining substances for toxicity by administering them to mice.

After 2,600 culture broths, three potential substances.

From May 1971 to January 1972, Endo's team investigated 2,600 culture broths. "We were doing grunt work every day until we got sick of it," Endo recalls. The lab was now cold and they taped the windows with packing tape to keep out drafts. They had found three strains of mold that passed the first five tests. But when they isolated the active substances, they found they had isolated oxalic acid and maleic acid. Endo says, "It had been known for a long time that many fungi produced organic acids, including oxalic, maleic, citric, and gluconic acids, and a similar level of inhibitory activity was recognized in all of these organic acids."

In other words, it was a dry run.

Endo recognized that their tests were too broad; they needed to eliminate these known active substances before running all the tests. Endo changed the team's procedure by using "extracts from which the organic acids had already been eliminated."

The Second Run Via an Improved Method

Throughout the spring of 1972, Endo and his team went through another 1,222 fungal strains using the improved method that eliminated organic acids. They found four strains that had inhibitory activity, of which two seemed promising.

One of these came from parasites found on rice and cucumbers. It turned out to be a known substance—citrinin. Endo says, "Citrinin was a strong inhibitor of HMGCR and, furthermore, when it was administered to rats for one week, a 14 percent reduction in serum cholesterol and a 36 percent reduction in neutral fat were observed. However, as previously known, kidney toxicity was observed and the research was halted. Although citrinin did not become the seed stock for a new medicine, this experience gave us hope and courage that we might be able to discover much better active substances in the near future."

That summer, Endo and his team began extracting and refining an active substance out of what they labeled "Pen 51." It came from a blue-green penicillium mold that came from rice found in a rice shop in Kyoto in the 1950s. Endo and his assistants found citrinin and other substances in the broth as well. To isolate and test the unknown active substance, they would need more broth, and by September they had cultured 250 liters of the cloudy mixture. That fall, using organic solvent extraction and chromatography, they isolated 8 mg of the active substance—a quantity sufficient for testing. The substance proved novel, and they named it ML236C.

Tests confirmed its strong inhibitory action on the HMGCR enzyme. Endo says, "It was recognized as a promising substance with inhibitory activity 1,400 times stronger than oxalic acid and 300 times stronger than citrinin. Needless to say, this was very encouraging for us."

Endo and his team kept growing more of the blue mold, and by February of 1973 they had 162 mg of the purified substance. When tested on mice it showed low toxicity. The next step was to test it on rats, but to perform adequate tests they needed 5 grams. At their current rate of production, that would require refining 156,000 liters of culture solution, about the size of a thirty-foot by thirty-foot swimming pool—an impossibility.

While the team continued its regimen of examining other fungi (with 2,570 more possibilities chosen), they confronted the entirely new problem of obtaining mass amounts of their promising candidate. They would need to engineer a way to increase the rate of fermentation of the mold. By modifying the incubation temperature, altering the ventilation volume during incubation, and changing the medium to 2 percent glucose, 2 percent malt extract, and .1 percent meat juices (peptone), they were able to increase the production rate by a factor of 40. Most of the improvement came from the addition of the malt extract.

By that fall, they were brewing great quantities of the broth, and when they analyzed it they found two new active substances besides ML236C—ML236A and ML236B.

In October, the crystal structures of all three were determined by another Sankyo lab using X-ray spectrography. They were all quite similar. However, ML236B was found to have an HMGCR-inhibiting potency 10 times that of the first substance discovered, and 20 times that of the other new discovery. So, Endo and his group decided they would develop ML236B.

By now they had finished analyzing the other 2,570 strains of fungi. Three of those molds showed strong inhibitory action, but none showed sufficient reproducibility to be considered promising as drugs. In two-and-a-half years they had analyzed 6,392 strains of fungi. Would the single isolated substance left, ML236B, later to be named compactin, prove useful?

The Cholesterol of Rats Does Not Fall Down

Some people ascribe the mental qualities of open-mindedness, nimbleness, even childishness to scientists. Scientists themselves often use the term "playfulness" to describe their approach. When confronted with these descriptions, Endo thinks of one particular playground toy to describe his experiences in science—the seesaw. By now, Endo felt like a kid riding a seesaw on the rise, looking up at the clouds floating by, anticipating the instant of weightlessness just ahead when the seesaw would fling him upward toward the blue sky of success.

Discovering a new drug in medicine differs from discoveries in other sciences in that the work is only half done when the discovery is made. Most of the discovery work occurs in test tubes, so any new drug must be proven to work in real living organisms. After that, it has to work in humans. And, finally, it must be safe.

In January of 1974, Endo sent five grams of compactin to be tested for toxicity and efficacy at Sankyo's Central Research Laboratory. In the interim, Endo and his researchers waited anxiously, but confidently. The tests were performed on young rats, the main test animal used all over the world.

On February 25, 1974, Endo was told the tests were complete. He was informed that there were no toxicity problems, but that compactin had not done what it was purported to do. It had not lowered the cholesterol levels in the rats at all. Endo reeled unsteadily and could not believe what he was told. He said, "The next moment I was struck with deep sorrow that our research of the past three years may have amounted to nothing."

The next two years were difficult for Endo—his life's work seemed to be on the line. If that work was going to be revived, it was completely up to Endo's lab to do so for everyone else at Sankyo was in favor of abandoning compactin. As the months went by, Endo ran through scenario after scenario. There was no question in his mind that the studies had been performed correctly, so why had the rats' cholesterol not been lowered? "It was a basic assumption at that time," Endo recalls, "that what would not work on rats also would not work on humans. However, after a short time I started to have doubts about this basic assumption." Searching the literature, Endo learned that another cholesterol-lowering drug worked on humans and dogs, but also did not work on rats. Endo says, "This gave us a ray of hope."

He had a new idea: "I doubted whether I should use young rats to evaluate compactin, even though it was the trend in those days. Cholesterol must be an essential ingredient for young rats growing up. There is no unnecessary cholesterol at all for them. So, when we give force to lower the cholesterol, they must resist it desperately. It might be natural that it is not effective for young rats. Therefore, if we could get old rats which have a surfeit of cholesterol, compactin might be effective for them. The patients to whom we administer compactin are often people who are advanced in years and who have a surfeit of cholesterol."

Middle-aged rats were not typically used for evaluating lipid-lowering agents, so, starting completely from scratch, Endo set up a room in the lab to raise his own. While he and his associates confirmed that compactin did not reduce the cholesterol of young rats at all, they found it did reduce the cholesterol of middle-aged rats by 20 to 30 percent for the first eight hours. But after that, even if it was re-administered, compactin had no effect on cholesterol levels. Stumped, Endo tried to understand what was happening inside the rats. Finally, after ruling out that compactin left the body or lost its inhibitory effect, he and his team determined that when rats were given compactin their cholesterol feedback system produced eight to ten times as much of the HMGCR enzyme. Thus, the more compactin given to the rats, the more cholesterol was synthesized, replacing all that was lost. The rats' cholesterol feedback system more than compensated for any HMGCR inhibition.

By January of 1976, two years after their triumphant discovery, Endo and his team finally understood why compactin did not work in rats, but they were no closer to creating a commercial drug. In fact, it was possible that the whole idea of inhibiting the HMGCR enzyme simply would not work because the body so desperately needs cholesterol, and would always compensate for any inhibition. Their five years of work looked fruitless; the prospects for compactin becoming a drug seemed remote. It was time to face giving up.

On the way home from the lab one day that dark January, Endo stopped at Yomogi, a restaurant and bar close to the train station. It was so small, with two tables and a capacity of around ten, that he couldn't avoid acknowledging the other customers. At the bar he recognized another Sankyo associate, Noritoshi Kitano. "I knew that Kitano was a researcher, a veterinary surgeon, at the Central Research Laboratory," Endo says, "but it was my first time exchanging greetings with him."

The very next week, Endo again by chance ran into Kitano at the restaurant. Endo says, "We engaged in a lively conversation, to some extent due to the influence of drink, and I learned that Kitano was conducting pathological research using laying hens. At that moment, I became aware of the fact that laying hens on a daily basis laid eggs which contained a large amount of cholesterol."

Endo, his mind ever at work, quickly realized that if the eggs hens lay are so rich in cholesterol, it meant they had to synthesize a lot of it. That meant they should have a lot of cholesterol in their blood, just like humans.

"Kitano told me," Endo says, "that he was going to clean up [kill] the laying hens after the examination, which would end in the middle of February. I asked him, after I told him about the plight of compactin,

'Could you administer compactin to the laying hens before you clean them up?' Kitano, who enjoyed drinking and had a pleasant nature, promised his cooperation on the spot." There was only one condition, Kitano told him, laughing: Endo had to take the chickens off his hands.

Endo readily agreed. "Because I took care of about 20 cocks when I was a junior high school student," Endo says, "I was familiar with not only taking care of chickens, but also the characteristic smell of a henhouse and the loud screaming noise of chickens."

Says Endo, "We divided ten white leghorn chickens into two equal groups. One was a medication group and the other was a control group. We fed commercial feed to the control group and commercial feed added with 0.2 percent of compactin to the medication group. My prediction hit the nail on the head, and within a period of two to four weeks the blood cholesterol of the hen fell by close to 50 percent. During this period, a reduction of more than 10 percent was observed in the yolk cholesterol. The laying hens stayed healthy during the administration of the compactin and after the administration was halted no pathological abnormality was observed. My team and I shouted with joy and toasted our success when we discovered the dramatic effect of compactin!"

In August, Sankyo put compactin, the first of a class of drugs now known as statins, back on track to become a drug. The seesaw had risen.

But, when the cherry trees bloomed in the spring of 1977, Endo could feel the seesaw slamming into the ground again. Research moved to dogs, which received different doses of compactin. The studies proved it to be completely safe up through doses of 250 mg of compactin per 1/kg of a dog's body weight, but at 500, 1,000, and 2,000 mg/kg, some of the dogs' liver cells developed minute crystals. This did not alarm Endo because he predicted the dosage for humans would be around 1 mg/kg. However, upper management at Sankyo wanted only risk-free drugs, and the discontinuation of compactin again seemed inevitable.

A Clandestine Human Trial

Most Japanese doctors were not desperate for Endo's drug. After all, lowering cholesterol for the average person with high cholesterol has little urgency, since it will only prevent heart attacks many years later. However, even in Japan, where few patients had high cholesterol, a few patients suffered from a rare form of cholesterol-related disease known as FH (*familial hypercholesterolemia*), with devastating symptoms.

Dr. Akira Yamamoto heard Endo speak about compactin at the annual meeting of the Japanese Conference on the Biochemistry of Lipids in Kyoto in June, 1977. Yamamoto had an 18-year-old patient with homozygous FH who had severe symptoms of blocked arteries, cholesterol deposits in her skin, and angina. She could not live a normal life and, without treatment, would almost certainly die young. The only way to lower her cholesterol thus far had been by intravenous nutritional therapy or by continuous feeding through a duodenal tube, both attempts to bypass normal food digestion that would pull fat and cholesterol into the body.

Desperate for a better treatment, Yamamoto, of Osaka University Hospital, buttonholed Endo. "My intention was not to test the efficacy of the drug, but to cure the patient," he later said.

Endo discussed the proposal with his supervisor, Arima. It was the second request they had received for a sample to test on patients in crisis. That same year, Sankyo had refused an earlier similar request from American doctors and researchers Michael Brown and Joseph Goldstein. Throughout the fall, Yamamoto prevailed upon Endo for a sample. Endo was an advocate, but Amira waited, hoping good news would come from Sankyo's other departments. Instead, the news grew steadily worse. Endo was told his presence at meetings about compactin were unnecessary; other departments at Sankyo were trying to synthesize compactin mimics, and so were in competition with the fermentation lab where Endo worked. If Endo wasn't around, their drug research might get more attention. Although Endo refused to be pushed aside and attended the meetings anyway, it was made clear that he wasn't welcome.

At the end of the year, evidence came out that Sankyo was going to kill compactin, so Arima decided to provide a sample of compactin to Yamamoto. Arima whispered to Endo, "We should not tell anybody about this for a while."

Endo himself carried 50 g of compactin to Osaka, where he met with Yamamoto, who introduced him to Professor Nishikawa. The professor told him that the hospital would take full responsibility for the human testing, and Yamamoto said that he would test it on himself. Endo bowed, then handed Yamamoto the small bottle of powdered compactin. Thus began a clandestine month of late-night telephone calls that originated only from Endo's home to avoid any overheard conversations.

The first patient treated was the 18-year-old girl. The drug was effective, dropping her serum cholesterol by 20 percent. But after three weeks she had muscle pain so severe that she couldn't walk, so the drug was stopped. Endo held on as the seesaw swung down again. There were, however, some positive signs. A test which listened to the noise in her carotid arteries (a means of discerning blocked arteries) indicated less blockage, and her skin xanthomas, which are fatty deposits caused by too much cholesterol, had begun to recede. Other doctors at the hospital objected to continuing the treatment, but Yamamoto persisted. A couple of days later the patient had improved, and Yamamoto lowered her dose from 500 mg to 200 mg. He also gave small doses to eight other patients for several months. Their cholesterol fell by 30 percent on average.

When Endo divulged the results of the trial to Sankyo, they could not deny the positive evidence and testing of compactin resumed. By that fall they had begun Phase I clinical testing.

Compactin seemed on track to become a commercial drug. It was a natural substance, and most people did not exhibit the side effects the first patient had. A later editorial in the *New England Journal of Medicine* said, "Compactin seems to have been designed by nature to be an ideal competitive inhibitor of HMG-CoA reductase. The enzyme has a 10,000-fold higher affinity for compactin than it has for the structurally similar substrate HMG-CoA."

Japanese Corporate Ostracism

With the clinical testing begun, Endo thought he could finally relax and turn toward new horizons. He moved to Tokyo University of Agricultural and Technology. "It was my dream for some time to move to a university once the development of compactin had reached a certain stage, and to conduct research together with young researchers."

Endo collaborated with Arima and Keisuke Murayama, the recently appointed director of the central laboratory. Endo says they came up with a plan "to send researchers to Tokyo University of Agriculture and Technology from Sankyo and continue collaboration with me about the theme of development of a cure for hyperlipidemia and obesity."

That all changed when word of Endo's retirement reached the top brass at Sankyo. At that time, Japanese corporate culture was very conservative. Many corporations viewed anyone who halfway resigned (the Japanese term for retiring by one's own choosing before the traditional retirement age of 55) as disloyal. Endo says, "I had wished that I would not have received the cold shoulder from my company because of the significant contribution I had made in the development of statin. However, I was treated exactly in the same way as my fellow workers who had resigned before me. My company's prohibiting my fellow workers from helping me clear my belongings from the office was just one example of this ostracism."

The lonely Endo carried his own boxes out of his office after working 21 years and eight months for Sankyo. His retirement allowance came to about $2,000 a month, the same as any other worker of a comparable position. As Endo says, he "never received any reward from Sankyo for the development of statin."

I Doubted an Ear for an Instant

The commercialization of statin drugs makes for a great case study in two entirely different business school courses. How a pharmaceutical company, Merck, obtained its version of the bestselling, most profitable drug in history would be a great study in how to maximize profits and completely change an industry. It also would be a great study in business ethics.

As Endo tells the story, "In April 1976, the head office of Sankyo received a letter from Mr. H. B. Woodruff, the executive administrator of Merck [then Merck, Sharp and Dohme] Research Laboratories. In summary, the brief letter more or less said: We have read the announcement of your patent application for compactin. The biochemists at Merck Research Laboratories have shown a keen interest and appreciate the opportunity to make an evaluation. If you are able to send a sample (5 g, if not possible, 0.5 g), we would be happy to forward a signed disclosure agreement." The letter from Woodruff also included the following leading statement: "We hope that as a result of these exchanges, a product will be found which is suitable for license and royalty return."

Over the next two years, Sankyo sent Merck much of its research. Endo says, "The attached materials revealed the structure of compactin, the fact that it was discovered from penicillium, has a competitive inhibition of HMGCR, is low in acute toxicity, significantly inhibits cholesterol synthesis in the cultured cells of both healthy individuals and FH homozygotes, and significantly reduces cholesterol in rats with hyperlipemia." Endo also visited Merck's research laboratories and consulted with them on how to test compactin. Sankyo delivered to Merck more than 100 g of compactin.

Then Merck became very quiet. Endo says, "We received no response from Merck and had a feeling of dread that maybe they were contriving a plot of some sort."

Finally, word arrived. Endo says, "In the middle of August 1978, we received word of the long-waited results: that the testing of compactin on dogs was extremely favorable. They contacted us by saying that they wanted us to supply toxicity data of compactin and a small additional supply of crystals. They also asked that they did not want us to engage in licensing negotiations with other companies while Merck was conducting evaluation tests of the compactin. From the wording of this letter, in addition to the relevant personnel at Sankyo head office, we also thought that perhaps Merck was seriously considering joint development of compactin."

In Japan's culture of honor, corporations often work together and trust each other. But in October 1978, Merck dropped a bombshell.

As soon as they had seen the first contract with Merck, Endo and Dr. Hiroshi Okazaki, his supervisor, warned the top brass at Sankyo that the written agreement providing Merck with samples of compactin was full of loopholes. They were providing Merck with an immense wealth of information. The discovery of penicillin had proved decades before that one great drug idea can lead to the development of many similar drugs. In the twenty years after penicillin was discovered, thousands of antibiotics were discovered, using similar techniques. Endo and Okazaki tried to get Sankyo to insist upon an agreement that would be common today—one that would not allow Merck to take the idea, modify it slightly, and then profit from it without paying a royalty. But the brass at Sankyo trusted Merck.

Merck informed Sankyo that it had tested compactin in combination with cholestyramine, another drug, and hinted that it might patent the combination as the invention of Merck alone.

Endo says, "The concomitant use test of compactin and cholestyramine would not have been a test for Merck to take the liberty to conduct under normal circumstances. Even if the test had been approved, the rights should have been shared between Sankyo and Merck. However, due to the details of the confidentiality agreement, Sankyo could not have made a complaint even if Merck kept the rights to itself."

Then, completely full of aplomb, Merck asked for 300 g more of compactin crystals.

Endo felt as though his stomach were falling through his feet. Here he was in the process of leaving Sankyo, and rather than reaping rewards for two decades' worth of discoveries, he was being tormented for doing so; and now he was learning that his drug might be commercialized without Sankyo getting any monetary benefit for all his years of work on it.

But just as in the darkest days of his youth, when the war was lost and "even kids lost their dream," Endo was able to summon strength. Resilient, he made the move to Tokyo University. "I prepared to start researching right after arrival to the post," he says. "The research was for the students, so I decided two things about the theme. First, that research was technically easy. Secondly, it did not cost much. One of the research projects was production research of *Monascus purpureus*."

Within a month, his research yielded another statin. *Monascus purpureus*, a mold found in red yeast rice (red koji), was found to contain a substance that was similar to compactin. Endo named it monacolin K. On February 20, 1979, he filed for a patent on it.

Later that year, Endo submitted an article about monacolin K to an academic journal. Then on September 11, 1979, right after the article was published, Boyd Woodruff of Merck showed up at Endo's laboratory. Endo relates that "he said, 'I read the article on monacolin K. Monacolin K is very similar to what the researchers of Merck have discovered. I would like you to offer us a small amount of crystal of monacolin K to identify whether they are the same substance.' So I doubted an ear for an instant." ("Doubting an ear" is a Japanese saying meaning "I can't believe my ears!") Endo was stunned. Merck had already said that if it developed compactin, it would likely do so in combination with another drug, circumventing any royalty payments. Now Merck was claiming that they had already invented the drug Endo had been first to patent.

Endo was a man of honor. He composed himself. "I took heart and handed him a few milligrams of crystal."

A month later Merck, wrote to tell Endo that its tests indicated it had already discovered monacolin K, before Endo had discovered it, although they had not applied for a patent until four months after Endo had applied for his. Merck had named it lovastatin.

Patent law differs between countries. Japan is one of about thirty countries where the date on which the first patent is filed takes precedence over the date of discovery. However, in the United States and some other countries, the date of discovery is deemed more important than when a patent is filed. In the end, Merck did not commercialize compactin. But it did commercialize lovastatin in 1987 and profited from selling it as the first commercial statin drug in countries that had favorable patent laws, under the brand name Mevacor.

Endo sums it up: "Merck signed the nondisclosure agreement, and they not only got secret information/documents and crystals of compactin over and over again, but also received our guidance for more than two years. Furthermore, they discovered mevinolin [lovastatin] without permission and have monopolized the right. It is as if we believed the proposal and dated for more than two years, but were betrayed. It will never happen in our country; however, we cannot complain because it was not a breach of contract. Still, it should not have happened if we had been used to how to associate with overseas companies."

The patent of monacolin K in Japan was granted to Sankyo, which paid 35,000,000 yen (around $140,000) to Tokyo University. The rest, as they say, is history. It is a history that those teachers who believe in Ayn Rand's laissez-faire capitalism can teach in business schools and those who revere less the principle of selfishness can explore in ethics classes.

Statins Prove Cholesterol's Role in Atherosclerosis

The idea that cholesterol was a risk factor for coronary heart disease was controversial throughout the 1970s and 1980s. British researcher Gilbert R. Thompson recalls that the head of medicine at Hammersmith Hospital where he worked was "virulently anti-cholesterol," meaning he didn't think it was a major risk factor. John McMichael would tell anyone who asked, "This relationship between cholesterol and heart attack, might it not be that the heart attack put the cholesterol level up?" The joke among cardiologists who were against the cholesterol hypothesis was "that a low-fat diet doesn't make you live longer; it will just make your life seem longer."

The evidence that people with FH (*familial hypercholesterolemia*) had clogged arteries and heart attacks as early as childhood was based on too small a data set to suit many scientists. It would take large, long-term studies to prove that high cholesterol causes the slow buildup in the arteries—atherosclerosis—that leads to heart attacks and strokes. But such studies were impossible before there was a way to prevent high cholesterol without side effects. Endo's statins were the first drugs that had few side effects, and therefore could be given to people with elevated but not immediately dangerous levels of cholesterol. The breakthrough study was the Scandinavian Simvastatin Survival Study

(4S). It followed 4,444 patients with high cholesterol, some given a statin, others not, in a double-blind trial, lasting more than five years. Cardiologists all over the world awaited its outcome.

The chairman of the study, Michael Oliver, was from the camp that was skeptical of the LDL-lowering treatments, believing that they wouldn't work or that they would have dangerous side effects. Afterwards, he quoted British economist John Maynard Keynes: "When the facts change, I change my mind. What do you do?"

The lipid study provided significant evidence that lowering cholesterol saves lives. During the study, 438 patients died, representing 12 percent of those receiving placebos, versus only 8 percent on the statins. Those receiving a statin showed a 25 percent reduction in cholesterol, suffered 42 percent fewer coronary deaths, and had 30 percent fewer deaths overall. What's more, there was no increase in non-cardiovascular deaths, meaning that statins had not harmed anyone.

A similar study in Scotland using pravastatin, another statin drug developed by researchers at Sankyo, showed the same results. The researchers extrapolated: "It can be estimated that treating 1,000 middle-aged men with hypercholesterolemia and no evidence of a previous myocardial infarction with pravastatin for five years will result in fourteen fewer coronary angiograms, eight fewer revascularization procedures, twenty fewer nonfatal myocardial infarctions, seven fewer deaths from cardiovascular causes, and two fewer deaths from other causes than would be expected in the absence of treatment."

While there is still some controversy about the importance of lowering cholesterol, there is overwhelming evidence that statins are lifesaving drugs when given to patients with a high risk of having a heart attack.

In the 1970s, Michael Brown and Joseph Goldstein discovered that the liver removes cholesterol from the blood by way of LDL receptors. Endo's statin drugs were perfect for increasing LDL receptors. Endo's original hypothesis was that inhibiting the HMGCR enzyme would lower blood cholesterol. While true, it turns out that it only does so indirectly. Statins inhibit the synthesis of cholesterol in the liver, which causes the LDL receptor genes in the liver to become more active. As a result, the liver expresses more LDL receptors, which pull more LDLs out of the blood, lowering blood cholesterol levels. There is some indication that statins also inhibit other enzymes that damage blood vessels.

In 1985, before statins were commercialized, Brown and Goldstein received the Nobel Prize for their work on LDL receptors. In fact, thirteen scientists who have devoted major parts of their careers to the study of cholesterol have won the Nobel Prize. To Endo, there must have been a sense of irony in the announcement of the award to Brown and Goldstein. The Nobel Committee's press release praised Brown and Goldstein for finding new treatments for high cholesterol—with statins. If you visualize a seesaw with those on one end describing a medical problem and those on the other end fixing the same problem, you would see thirteen scientists laughing in childish glee as they rise through the air, feet flying, having the time of their lives accepting their Nobel Prizes. On the other end, down on the ground, would be Endo. It is as if the person who describes a problem gets all the credit, while the person who solves the problem is forgotten.

Brown and Goldstein are longtime fans of Endo and have said, "The millions of people whose lives will be extended through statin therapy owe it all to Akira Endo and his search through fungal extracts at the Sankyo Company."

Big Pharma

Endo and his team knew the molecular shape of compactin thanks to tests such as Nuclear Magnetic Resonance (NMR) and mass spectrography, amongst others. The shape consists of four parts: a head, a neck, a trunk, and an arm. Their analysis had shown that the head is the part that binds to and inhibits the HMGCR enzyme. They knew that changing the arm part could limit the inhibition because when they initially discovered the three similar compounds, the main difference between them was in the shape of the arms. Drug companies had the tools to examine Endo's statins with just such precision. Soon, companies all over the world were trying to synthesize statins, tweaking them just enough to obtain a patent. They typically left the head and neck part alone, and replaced the trunk part or the arm part. Novartis developed fluvastatin; Warner-Lambert developed atrovastatin; and Pfizer developed atorvastatin (Lipitor). Sankyo and Merck each developed lovastatin, calling them different names due to the patent disagreement. Continued research allowed Merck to get around the patent problems that prevented lovastatin from being sold in all countries. It synthesized another statin—simvastatin—and named it Zocor.

Statin drugs were so popular that they transformed drug companies into behemoth corporations. Profits could go not only to shareholders, but also into research laboratories, which churned out still more life-saving drugs. In 2005, of 10,000 drugs sold in the world, the best- and fifth-best-selling drugs were Lipitor with $12 billion in sales and Zocor with $4 billion in sales. Merck's stock price more than doubled in the seven years after it introduced the first commercial statin. In the six years after the Scandanavian 4S study came out, the share price of its stock quadrupled.

Sankyo also developed different statin drugs, putting the company on a par with other international drug companies. It was so important to the corporation that when Sankyo built its main office in Nihonbashi, it named it the "Mevalotin Building," after one of its statin drugs. Worldwide sales of statin drugs have far surpassed $100 billion. Endo, ever riding the seesaw, never received any money from Sankyo or any other corporation for any statin drug. Japanese ostracism has a long memory. Endo's name is not mentioned on Sankyo's website to this day.

Statin drugs only became widely prescribed in the late 1990s. Their full effect is years, perhaps decades, in coming, so they haven't yet saved the number of lives that other discoveries in this book have. Even so, Amy Pearce estimates that they have already saved more than five million lives from death by heart attack or stroke.

"Statins are to cardiovascular disease what penicillin was to infectious disease," says lipid expert Professor Leon Simons, head of the cholesterol clinic at Sydney, Australia's St. Vincent's Hospital. "They are one of the most important, if not the most important, advances in cardiovascular medicine."

Currently, 10 percent of the adult population in the United States takes a statin. Moreover, 27 percent of those older than 65 swallow the precious pills every day. While this calculates out to around 20 million individuals, it is estimated that twice as many should be on statins. More recent studies have further substantiated the value of taking statins. They even give hints that the drugs may prevent various other diseases, including some forms of cancer. Some advocates have gone as far as suggesting that we lace our food or water with statins, the way we add folic acid to bread and fluoride to drinking water. Others believe they should be sold over the counter. These moves may not occur, but they indicate how powerful these drugs may end up being.

Today, millions of World War II veterans extend their lives with statins, a drug discovered by a child of the new Japan they helped create. Not only has Endo received no money for his discovery, but virtually none of the old soldiers know his name.

Like a Seesaw

Endo and his wife raised three children. Today he is retired, living in Tokyo. He still climbs Mt. Chokai, a grand mountain near his hometown, with his wife and his old friend from junior high school, Hatakeyama. Endo says, "Now at over 70 years of age, I still enjoy trekking in the mountains around Tokyo, and my foremost joy is listening to the music of Beethoven, Mozart, and Bach."

Through it all, Endo has kept a good humor. In 2004, a Festschrift (a German word meaning "celebration publication") was held in Japan to honor Endo. Many relevant scientific articles and memoirs of people with whom he had associated were put together and published, and a dinner was held in his honor. When it was Endo's turn to speak he said, "Today I have good news and bad news. The bad news is that my cholesterol levels have reached 240 mg/dl. Maybe I had sukiyaki or shabushabu too often! The funny thing is that the doctor said, 'Don't worry! I know some very good drugs to lower your cholesterol.'"

Obviously, the doctor did not know the drug's creator.

Through the years, Endo has become philosophical about his lack of money. He says, "The discovery of the medicine was a continuation of difficulty which was like a seesaw. However, when I pass the difficulty, there is the pleasant feeling like after I exercise and sweat. The discovery and development of statins was a huge gamble for me in achieving my dream. Thanks to my success with statins, the dream from my boyhood was realized and I received appreciation from a large number of people from all over the world. This is a source of immeasurable joy for me. During the development and research of statins, I had the opportunity to meet many outstanding scientists. Of these, I learned a great deal from three great scientists in particular: Konrad Bloch, Joseph Goldstein, and Michael Brown."

Endo recalls his goal beginning the research: "I did not start the research to make money or become a big man. Since I was born as a human in this world, I wanted to leave my mark before I die. I want to die after I do at least one thing useful for the world. I could start the research because I had such a thought. Therefore, we cannot measure the contribution which statins made toward saving precious lives. Maybe we should not simply convert 'to be useful for the world' into money. It is something we cannot convert."

Endo continues, "Nowadays, it is said that money is important. However, we can discover the pleasure and value of life when we do something for the world with a sense of mission. What I have done was rather for the world than a Japanese company or Japan. It was needed all over the world, so I challenged for it. Especially now, it is called the time of globalization, and borders are not clear. Humans made borders originally, so they exist, but they seem not to exist. I want to tell young people the message that the philosophy and sense of value of doing something for the world are more important than making money. That is the work left for me from now on."

Lives Saved: Over 5 Million

Discovery: Statin drugs

Crucial Contributors

Akira Endo: After analyzing 6,392 different molds and fungi, Endo discovered a class of drugs that became known as statins. Statin drugs lower LDL cholesterol in the blood, thus preventing atherosclerosis, a clogging of the arteries which can lead to heart attacks and strokes.

Akira Endo
in the lab,
pictured
soon after
graduating
college.
Photo courtesy
of Akira Endo.

Endo was eleven years old when the war ended and recalls, "We lost the war. Even adults did not know what to do. We did not have food and clothes. Everything was called off. Even kids lost their dream. It was so hard that even kids did not have hope."

Tomiko, Endo's sister-in-law, recalls the trip to see Endo. "His parents asked us to tell him that it was 'not only you, you have brothers and sisters, so you cannot go to a university.' They gave us 20,000 yen [about $55] and said to us, 'tell Akira if he does not go to the university, we will buy him a suit. Trick him.'"

"Since I was born as a human in this world, I want to leave my mark before I die. I want to die after I do at least one thing useful for the world."
—Akira Endo

Akira Endo, pictured around the time of discovering the first statin. Other than his normal salary, Endo never received any money from any drug company for his lifesaving discovery.
Photo courtesy of Akira Endo.

"With that said, there was no guarantee that I would be able to discover an HMGCR inhibitor, so I decided to study a few thousand strains of fungi over a two-year period. It was going to be a research like a bet...and if I was unable to discover the substance I set out to find, I would then terminate my research."
—Akira Endo

In two-and-a-half years, Endo's team analyzed 6,392 strains of fungi to discover one substance that might work. Recalls Endo, "My team and I shouted with joy and toasted our success when we discovered the dramatic effect of compactin!"
—Akira Endo

"I think there is no doubt that the discovery of compactin (the first statin) by Dr. Endo was one of the great discoveries in the history of modern medicine."
—Scott M. Grundy, MD, Ph.D., University of Texas

"Statins are to cardiovascular disease what penicillin was to infectious disease," says lipid expert Professor Leon Simons, head of the cholesterol clinic at Sydney, Australia's St Vincent's Hospital. "They are one of the most important, if not the most important, advances in cardiovascular medicine."

Akira Endo and students in 2007, collecting soil samples from which fungi are isolated and grown. Photo courtesy of Yutaka Nakamura.

Coronary heart disease caused 452,300 deaths in 2004 and is the single leading cause of death in America today for men and women. Over 15 million Americans have had a heart attack or angina. This year an estimated 1.2 million Americans will have a heart attack.

From 1994 to 2004, the death rate from coronary heart disease declined 33 percent, thanks to Endo's statins and other medical advances.

In 2005, out of 10,000 drugs sold in the world, the best-selling drug was a statin—Lipitor—with $8.4 billion in sales. Worldwide sales of statin drugs have far surpassed $100 billion.

Chapter 3: Bill Foege—
Over 122 Million Lives Saved

The Eradication of Smallpox

BILL FOEGE grasped an arm above the elbow, held it firmly, pressed the gun to the skin, angled it until it was still, pulled the trigger, and released the arm. Then he pressed a hydraulic pedal with his foot, and grabbed another arm. The black gun he used was the size of a 9mm Luger. A hose fell from it to the foot pedal, and a bottle was attached to the top, ensuring that what came out of the gun saved lives, rather than spent them. It was 1964 and the 28 year-old Foege (pronounced FAY-ghee) was on the South Pacific island of Tonga. He was vaccinating a long line of people with the Ped-O-Jet gun, a jet injector that delivered doses of vaccine. He tried to get into a rhythm because when he found a cadence he didn't lacerate arms by shooting too early, plus it went faster.

Men, women, and children moved past him, each receiving an injection of 1/300th of an ounce of vaccine into the space just beneath their skin, where the cowpox virus would begin multiplying. Foege had been going through the routine for several hours by now and as he tired, the repetition became drudgery. He didn't mind vaccinating people—it was what he had been trained to do; it was his duty. Naturally, in medical school he had dreamed bigger, with a young person's visions of curing rare tropical diseases, or performing groundbreaking medical sleuthing.

But it was vaccination, he knew, that was the key to preventing smallpox. Some even held the audacious view that smallpox could be eradicated. It was certainly a disease worthy of elimination. Smallpox had destroyed the Incas, the Aztecs, and other American Indian nations, killing as much as 95 percent of their populations. It had raged through Europe and America in the 1800s. Even in 1964, it was still striking an estimated ten to fifteen million people a year in forty-three countries, killing more than two million. In thirty-three of those countries it was endemic, continually present in the population.

Smallpox is a poxvirus, one of a family of related diseases named after the creatures most likely to contract them: camelpox, raccoonpox, mousepox, monkeypox. There are two clinical forms of smallpox—Variola minor, which is the less lethal form, but which scars many of its victims; and Variola major, the most common form, and the big killer. It slays 30 percent of its victims, blinds 10 percent of the survivors, and scars almost everyone. The viruses, which sometimes look like a dumbbell under an electron microscope, are among the largest and most complicated known.

Yet the disease is entirely preventable. Long before, in 1796, Edward Jenner had performed experiments that demonstrated that getting cowpox—a mild form of poxvirus common in cattle—immunized a person against smallpox. Jenner extracted cowpox and named it vaccinia ("vacca" being Latin for cow), and then he deliberately gave it to people. Thus vaccination was born. Jenner's cowpox virus had been passed down all the way to Foege, mutating along the way into a slightly different virus, but still making whoever received it immune to its killer cousin, the smallpox virus.

Nevertheless, many experts thought it was impossible to eradicate a disease from the entire earth. After all, it had never been done before. How can you go after a germ as small as a virus and destroy every single one on earth? The famous microbiologist René Dubos wrote at that time: "Eradication of microbial disease is a 'will-o-the-wisp;' pursuing it leads into a morass of hazy biological concepts and half-truths." It would take vaccinating the masses of humanity to eradicate a disease, but how do you freeze people in place so that you can walk up to them and give them a vaccination? Children are constantly being born; nomadic people are moving around; a small percent of any population keeps its distance from the rest—itinerant workers and street people, criminals and drifters, all fervently fleeing authority figures, furtively existing underground. In addition, some countries have political leaders who have no concern for their people and so refuse vaccination programs, and in every country there are groups who fear or don't believe in vaccination itself.

However, there are always the cockeyed idealists, awakening every morning to blue sky and sunshine, believing that science inexorably improves civilization. They argued smallpox was eradicable because it infects no other species, only humans. If for a single two-month period no one on earth caught smallpox, then the disease would either kill or run its course in whatever humans were already infected, and that would be it—the virus would cease to exist. This had happened in North America and Western Europe, so why not the whole world? The key in the smallpox-free areas was mass vaccination. So many people had received vaccinations that the virus had no one to spread to and, poof, it was gone. Virologists call this herd immunity. If the whole herd of humanity were immunized, the virus would die. A 1964 World Health Organization advisory committee concluded that vaccinating 80 percent of a population would fail to eradicate smallpox. It recommended 100 percent mass vaccination.

Foege sighed. Certainly this rote vaccinating prevented disease, but it was very boring work. His job was to be a robot, endlessly injecting people, only occasionally stopping to fix the gun. He would set his record in a prison. Long lines of expressionless men shuffling through the sand, prodded by sticks to keep moving, added up to 600 men vaccinated in just a bit over 20 minutes. His one-day record, set on one very long day, was 11,000 people vaccinated.

The Son of a Lutheran Minister

If anyone was an idealist, it was Bill Foege. His father and grandfather were Lutheran ministers and his uncle had been a missionary to New Guinea. At age 15, an injury put him in a body cast for three months. Immobile, he could do nothing but read. He read the books of Albert Schweitzer and imagined a life in the jungles of Africa. One Schweitzer quote in particular remained with him: "No one has the right to take for granted his own advantages over others in health, in talents, in ability, in success, in a happy childhood or congenial home conditions. One must pay a price for all these boons. What one owes in return is a special responsibility for other lives."

Foege grew up in the Pacific Northwest to be a rail thin, unintimidating six-foot-seven-inch lover of humankind. His sister Grace went to medical school and impressed him: "Weekend visits to her world revealed people totally captivated by a medical environment that few others knew existed." He followed her path through

Pacific Lutheran University in Tacoma, and then married his sweet-heart, Paula. After medical school at the University of Washington, he went into the Epidemic Intelligence Service at the Centers for Disease Control, which sent him to India with the Peace Corps, and then to Tonga in the South Pacific.

The herd of people kept moving, halting, moving. The Ped-O-Jet was now sweaty in his grip. The tiny arms of children, the cool arms of women, and the strong arms of working men passed steadily by. Press the foot pedal, grab another arm—vaccinating every single person on earth was going to take a while.

Foege's mind wandered, and he began trying to create and solve mind problems to keep up his energy. Could he visualize all the steps involved in creating this vaccination he was giving? He thought, "Every step involved in the basic research, the appropriation of funds, the deliberations of Congress, the university structure for the research, the corporations and their work in producing the vaccine, the packaging, cardboard, glass, rubber stoppers, needles, syringes...things required to put out a vaccine product...these jet injectors developed by Aaron Ismach for the U.S. military that could provide uniformly high take rates...the airline industry, the customs infrastructure that would inspect the vaccine, the Land Rovers, vaccinators, health educators...the number of teachers who had influenced every person in the chain of vaccine transmission...millions of people involved in the million steps...."

Ahead of him there was still a long line of people, many of them dressed up just for this day, to be vaccinated by the vaccinator, men with their hats, women and girls in bright colored wraps.

"...And at the end I was simply pulling the trigger on a jet injector. Why that should be recognized anymore than the thousands of steps along the way is a mystery. The miracle was that every one of those million steps had gone right or the vaccine could not have been given. That is the essence of globalization. Everything working for the benefit of an individual."

A child hesitated, gently pushed onward by her mother. A tear of fear arced down her round cheek, big brown eyes staring up at Foege—was he going to hurt her? He smiled, knowing it would not hurt. She held out her little arm in unknowing trust. Holding that arm, giving the injection, knowing from everything learned in church, everything learned in medical school, everything learned working a part-time job from his boss, Rei Ravenholt, or from Charlie Houston (good old Charlie, who was almost the first to climb K2 but came down instead, trying to save the life of a fellow climber), who in India taught him to get out of the office and to visit hospitals, get to know the real people in their real environment. He knew that he was in the right place, doing the right thing—vaccinating the herd.

Following in Schweitzer's Footsteps

When his Tonga tour ended, a commencement speech he had read inspired Foege to go back to school. In the speech, Tom Weller seemed to speak directly to Foege, who paraphrased it frequently: "You're only going through life once—you might as well try to do it right—and if you're getting all this training, skills, knowledge, and so forth, you should ask where could it do the most good." Foege immediately applied to Harvard, and nowhere else. He got in and, mentored by Weller, received a Masters Degree in Public Health. He came out as an epidemiologist, and even more of an idealist, believing more strongly than ever in global health.

Still under Schweitzer's spell, Foege, along with his wife Paula and their infant son David, boarded the *Aurel*—the flagship of the Elder Dempster Line—in Liverpool, England. They were off to Africa to work in a Christian mission. Over the next few days the ocean liner sailed south. Their first signs of Africa were high yellow dust clouds wafting out from the Sahara. As they approached a port, the lush smells of the continent awakened their senses. Passengers bedecked in beautifully colored, shimmering silk clothes boarded to travel to the next port, sometimes singing and dancing on deck. At a stop in Liberia, the ship picked up President Tubman, on his way to the African Unity meeting. Foege remembers a different Africa in those days: "Africa was a heady place. Everyone had the sense that they were on the way up. They had shaken off the suppressors—the European powers—and they would now show the world what they would do. Everyone felt good and almost giddy about the future."

The ship turned east toward the Bight of Benin, and then coasted to a stop in Lagos, Nigeria. The Foeges departed and drove inland

451 kilometers to Enugu, the capital of the region, then another 140 kilometers east to Yahe, where the Immanuel Medical Center was located. The Lutherans had been in Nigeria since 1936 and had set up the clinic to serve the surrounding villages, which had a combined population of about 150,000. It was a small clinic, consisting of a few rooms and a laboratory.

The next day they bumped over dirt roads to a village. The majority tribe in eastern Nigeria were the Ibos, but around Yahe there were numerous minority tribes and the clinic needed the Foeges to learn a minority language, Yala, which is tonal and unwritten. The villagers were primarily farmers who raised a diverse variety of produce, including black-eyed peas, peanuts, yams, taro, and cassava, a shrub with starchy roots. The Foeges moved into a mud-walled house with a metal roof.

There was no electricity or indoor bathroom, but they had a kitchen, two bedrooms, and a living room. Their African neighbors were known to be friendly, and quickly proved it. It was the custom in the tribe that all living rooms were a sort of community property, so people freely dropped in to greet, meet, and watch the giant American and his family. It was completely safe in the village and Foege had no qualms about leaving his wife and son if he was away for a night on a medical emergency, but the villagers insisted on providing a night watchman, whether Foege was away or not. It was never lonely in that village.

Foege's work as a doctor at the clinic reinforced what he had learned in India, "that most people in the world do not present a single medical problem. Whatever brings them to the hospital or clinic brings with it a background of malnutrition, roundworms, hookworms, and a half-dozen other conditions."

As they began to acclimate to African village life, Foege and his wife realized their lives would never be equal with those of true villagers. He says, "In one sense, it was an opportunity to know something about what it's like to live in a village in Africa. On the other hand, I've always said we could never truly feel that because we could leave anytime we wanted to, anytime it got to be too much. We were able to, during the dry season, hire a young man on a bicycle who could go to the water hole five miles away and get water and bring it back and put it in a fifty-five gallon drum for us. So we were able to do things that a villager couldn't, and I've said that the big lesson for me was, number one, that here we were living in third-world conditions and yet our son was living with first-world risks, the reason being that we could apply all the knowledge we had on immunizations, on screening the windows, on boiling the water, and so forth. But, lesson number two, if you had limited me to a dollar a day, I would have to have spent that on food and shelter and could not have done these other things—could not have afforded vaccines or even firewood to boil drinking water, and so this combination of knowledge applied and poverty, those are the two things that separated us."

While Foege was far away in Nigeria, the Nineteenth World Health Assembly voted in May 1966 to institute a worldwide small-pox eradication campaign. Such an initiative had been suggested in the 1950s by the Russians when their delegates returned to United Nations programs after Stalin's death, but it took a decade to gain the financial cooperation of the rest of the world. Now both Russia and the United States agreed to provide funding. The United States Agency for International Development (USAID) agreed to provide commodities, vaccine, jet injectors, and vehicles for twenty countries in Western Africa, with the CDC coordinating the efforts. Each country's ministry of health would provide manpower.

The CDC sent Henry Gelfand, a prominent public health physician, to Africa to lay the groundwork for the eradication campaign. West Africa had the highest rates of smallpox on earth, and so would need strong direction. Gelfand had been prodded to look up a tall, clean-cut, practical joker who those at the CDC thought would be perfect as an advisor. Gelfand was skeptical of a "missionary guy," but reluctantly agreed to meet Foege, if only to placate the home office. It didn't take him long to recognize that Foege was highly intelligent, had wonderful people skills, and was full of energy.

As for Foege, how could he pass up the opportunity to rid the world of a pestilence of truly biblical proportions? So, in September of 1966, he began spending weekdays in Enugu at eastern Nigeria's Ministry of Health, returning to the mission compound to work weekends.

Ogoja

Sunday morning, December 4th, 1966, was sunny and warm. The fronds of the African oil palm trees were still except when a grey parrot lighted on a branch, searching for fruit. People walked back from church, their children running ahead. Foege was in the clinic when he was called to the radio.

He walked into the radio room, picked up the microphone, and spoke into it. A second later, static, then, "Bill, I think I have some cases of smallpox."

It was Hector Ottemiller, a missionary.

"Describe them for me, Hector."

"They've been sick, and now round lesions have appeared on their face."

"I better come check it out."

Foege told Paula he had to go. He put his medical gear into a leather shoulder bag, tucked himself into the mission's VW Beetle, and drove northeast through Ogoja Province. As the kilometers fell away, Sunday afternoon scenes of rural African villages passed his view: women carrying vegetables on their heads, children dancing. He passed Nfom, Okpudu, and Okuku before meeting up with Hector and his colleagues. Hector was savvy. He had everything they needed ready, knew exactly where they were going, and would have personal contacts all along the way. He handed Foege bikes he had borrowed, which they fastened outside. When they arrived in Alifokpa, where the road ended, they parked. It was twelve kilometers on to the village of Ovirpua, where the smallpox cases were, so they unloaded the Velosolex bikes. Made by a French firm, these bikes had a small engine that turned a ceramic roller on the front wheels.

Off they went into the African savannah. It was a great afternoon for a ride. They bumped along the dusty red path, snaking through the tree-dotted grassland. Sometimes they could see to the horizon, sometimes patches of dry grass over their heads hid the entire view. Decades earlier, they might have been worried about lurking lions or rhinos—older missionaries told such stories—but now the big game was gone, restricted to protected parks. They might see a snake, but nothing large. Foege's bike was small for his large frame. His knees stuck out, and on big bumps his legs sometimes went flailing. Steadily, they motored through the brush. When they came to a creek they got off, picked up the lightweight bikes, and carried them across a log acting as a bridge. Then it was back on the bikes, past Acadia trees, some with big, dark, brooding vultures weighing down their branches. Occasionally, they saw a baobab tree—the "upside down" tree, with its huge girth of bulbous trunk—its dry edible leaves hanging limply from a few stout branches. It was undoubtedly the dry season, and dust was clouding up and covering their faces. But the heat was quite tolerable.

They arrived in the village of forty mud-walled houses covered by drooping, thickly-thatched roofs. The charismatic Ottemiller started talking to the villagers in the Yache language. Foege had heard it at the clinic, but did not understand it well. They were led to a hut, and a woman came to the door. Ottemiller, translating, talked to her. She disappeared into the darkness. A couple of minutes later a man emerged into the light. Only a little younger than Foege, he was barebacked and barefoot. His shoulders were slumped and he moved with the hesitant gait of the sick. The woman followed him and placed a mat on the ground. The man sat down. It was immediately evident to Foege that the man had smallpox. There were between 100 and 200 pustules on his face, sweeping back to cover his ears, then down his neck and onto his arms. Foege set his brown leather shoulder bag on the ground and knelt down. He didn't need to look at the man further. The pustules were hard and were all at the same stage of development, two of the major criteria for diagnosing smallpox.

However, it seemed the doctorly thing to do to pretend to diagnose him, to spend a minute with him. The man's wife crouched in fear, hoping he had something other than smallpox, but knowing inside the truth. Foege had him open his mouth. It was full of sores. Foege asked him, through Ottemiller, if he could eat or drink, knowing that some people's mouths were so painfully sore they refused any nutrition. Then Foege confirmed to Ottemiller that the man had smallpox, and Ottemiller translated the news. The man's wife, still crouched barefoot in the dust, grasped her head with both hands and bent it down toward her knees, rocking back and forth, reactiing to the unwanted, tragic knowledge. There was no treatment.

Foege knew well the course the disease had taken, and what would follow. The virus had entered through the man's nose or mouth as he breathed in virulent droplets from an infected person's cough or sneeze. A week, possibly two, had passed with no symptoms, as the virus incubated in the man's cells. Then his body began to ache, maybe he vomited, and then he developed a high fever of 101 to 104 degrees F. At this point he might have thought he had the flu. But then a rash with small red dots had developed in his mouth and on his tongue. The sores had grown and broken up, spreading the virus further into his mouth, beginning the most contagious phase. Then the fever had subsided and the man probably thought he was getting better. But the rash spread to his face, then down his arms and legs, and onto his hands and feet. By the third day of the rash, the red dots had started rising. Maybe he had chickenpox? The fever returned. Now the man's bumps were turning into pustules, sharply raised, a small indentation like a belly button in the middle, firm and round. To the touch, it felt as though BB pellets lay beneath them. It wasn't chickenpox.

The next couple of weeks would be horrible. In addition to the fever, the pustules would remain, slowly scabbing over. If the man lived, most of his scabs would fall off during the third week, leaving pock marks, scars for life. He would be contagious until they were all gone. If there were so many that they ran together so that they were confluent, survival chances were diminished. In Africa, the mortality rate was typically 20 to 25 percent.

Quickly they surveyed the village—there were three other smallpox cases. Foege had vaccine in his leather shoulder bag and as the sun lay low in the sky they began vaccinating the villagers.

That night, kerosene lights cast a ghostly glow out into the sky. David Thompson, also a young, idealistic doctor bent on saving the world, and also a future Christian missionary, and Paul Lichfield, a Mormon, had joined Foege. The three made up the smallpox eradication advisors to the Ministry of Health for eastern Nigeria. They sat around a table, trying to come up with a plan. They had radioed out and learned there was no extra vaccine, immediately thwarting the standard response they had been trained to put into effect when smallpox was found—mass vaccination. The eradication effort was not scheduled to begin until the next year, so ample supplies had not yet arrived.

The night sounds emanating from the darkness outside the radius of their lantern seemed to enhance their silence. Four smallpox cases had suddenly appeared. How many more cases were out there, lurking in dark huts? They had to face the grim possibility of an epidemic.

"What do we do with insufficient vaccine?" one lamented.

"What else can we learn about this outbreak?" another suggested.

They considered what they knew. One option was to dilute the vaccine they had on hand. This had been tried a decade before in Liberia. But they knew the take rate, the number of those who incurred an immune reaction when vaccinated, had been low. If they diluted their vaccine, many supposedly vaccinated people would be unprotected.

Perhaps their vaccine was stronger? Doubtful. They decided they would not dilute their preventative medicine. "If we were smallpox viruses, bent on immortality, what would we do next to survive?" Foege asked. They began imagining they were the virus. The obvious answer was to find someone else to infect, thereby ensuring reproduction.

How well did they know the virus? The first thing to do was find out where it was, but they were in the heart of Africa. Communication was poor. Not only were there no phones in many rural areas, even in the cities phone service was spotty—making a call out of the country required calling an operator and scheduling a sometimes elusive time days later to make the call, and then yelling to be heard.

But wait—the missionaries had other means of communication. Every night they used ham radios to check in with each other, to keep in touch, and to get assistance for any emergencies in their villages. There were fifteen missionaries in the surrounding area and Foege knew them all. The three eradication officers laid out maps of Ogoja, and when the radio crackled to life, they were relieved. They could now take their first action. They asked the other missionaries if they would help, and one after another agreed. The plan quickly coalesced: Each missionary would send someone, generally by bicycle, to the villages around them to record if any village people were sick with smallpox.

The next evening at 7 PM the missionaries reported back. In one day, they had surveyed a large area, each recording the smallpox incidence in two to seven villages (totaling almost a hundred villages), representing some 20-25,000 people. The missionaries had found six cases. The next day, vaccination teams were sent out to vaccinate the people close to each case.

The three advisors knew that smallpox was highly contagious, and feared that the cases found so far might be the first of a mushrooming epidemic, so they kept strategizing. Now that they knew where the virus had been several weeks earlier, when the sick had to have been exposed, could they draw its vector on a map and predict where it might go next?

This wasn't a straight forward geometry problem for Foege and his colleagues were unsure how smallpox was likely to travel between people. Talking with both missionaries and the villagers, they ferreted out two important means of transmission. The virus would most likely travel toward family members and marketplaces, where people in African village life connected.

Still trying to think like a smallpox virus, Foege and the others mapped out possible paths. Setting aside a supply of vaccine for the additional cases that might yet be reported, the team allotted some of the remaining vaccine to encircle the areas connected to the villages where the ten cases had been found. They chose the three most likely paths the virus might take, and vaccinated along them to a village marketplace and to two villages connected by family members.

Every night at 7 PM the missionaries reported in, fifteen earnest voices, broken by the staccato sound of static. Over the week, only four new cases were reported, and they began to feel more confident. During the second week an additional twelve cases were found—not so good. The third week brought nine new cases. What's more, the cases appeared in two of the three areas they had vaccinated. They had guessed right. Now they had to wait to see if they had reached the villages in time to stop the virus's spread throughout the populations.

During the fourth week, day after day they awaited the radio messages. One missionary after another reported no cases. With each day hope grew brighter. There was no epidemic—in fact, there were no more cases at all. Foege and his colleagues had stopped the virus.

As more weeks passed and additional vaccine supplies came in, they stepped up vaccination in the area. Then Foege had an interesting insight: "We might feel very good about every vaccination we were doing, but in the absence of the smallpox virus it was a wasted effort." The vaccines might be important when smallpox came through in later years, but in the short term they were throwing water on ground that was not on fire.

Foege had been trained in epidemiology, and was not about to let this data set go unexplored. He kept records and pored over the results. He calculated the number of people they had vaccinated in order to stop the first outbreak—7 percent. That was hardly mass vaccination. No herd immunity could have been conferred by what they had done. Foege enjoyed data and loved elucidating patterns. He believed in a cause-and-effect world. What they had accomplished was no fluke. There had to be a reason.

And the first man with smallpox Foege had seen in Ovirpua? He lived.

A Firefighter Discovers a 6 Percent Solution to a 100 Percent Problem

Over the following weeks, Foege devoured all the information he could find on smallpox. His colleagues knew him as a deep thinker with a phenomenal memory. One thing in particular he found curious was that the number of smallpox cases they had seen was small. Why was this, Foege wondered, when "folklore and the textbooks presented a disease of rapid transmission?" He explains, "The textbooks kept saying that this was one of the most contagious diseases. We believed them until we started the epidemiology. We went to villages where smallpox had broken out and found that there hadn't been a case for fifteen or twenty years. This is unlike measles, where every other year you expect an outbreak. So gradually we developed the idea that smallpox really comprises islands of infection that move around slowly, but is not something that covers an extensive area like a country all the time. During the smallpox season in West Africa, maybe only one percent of the villages contain cases at any one time."

Parasitic pathogens evolve clever tactics to survive. If they kill their hosts too rapidly, they will not have any place to live. It is therefore to their benefit to allow a host to survive long enough that they can multiply, and then jump to another host. When syphilis first appeared in Europe in the 1490s, it killed within months. But over time it evolved into a pathogen that took decades to kill. Some predict HIV will weaken and kill more slowly as time goes by, in order to optimize its survival. Smallpox had likely evolved to cause a long illness—up to two weeks with no symptoms, and then three or four weeks of sickness. Because it could reside in individuals over such a long period, it did not need to infect a large portion of the population at any one time. To survive, it only needed to pass itself on to two or three other people, and could do so slowly. Foege realized that "the tenacity of the virus in continuing to infect new generations within a household was confused with high transmissibility."

Foege had spent his college summers fighting fires in the Colville and Wallowa National Forests in the Pacific Northwest. His trainers had pounded into him and his fellow firefighters the principle that a fire can't burn without oxygen or fuel. He drew upon that analogy now. During forest fires, dirt was shoveled on flames to block oxygen, and firefighters built a perimeter of scraped dirt around fires to block access to fuel. Foege wondered if he and his colleagues could not do the same with smallpox. Susceptible people were the fuel of smallpox, and if they could vaccinate enough people, the virus would be snuffed out. They could also scrape away a perimeter of fuel around the islands of smallpox, leaving the virus to burn out, with nowhere to spread. In fact, the perimeter they needed to scrape only had to be a few feet wide. Studies would show that smallpox was usually transmitted due to close range (within six feet), prolonged contact, usually of more than seven days. Foege realized, "It was necessary, of course, to prevent the virus from traveling out of that protection bubble on contaminated clothes, as when a fire crosses a fire line, but the basics for breaking transmission of the virus were remarkably simple and similar to fighting forest fires."

Foege had in mind a radically different strategy to deal with smallpox. If smallpox was not ferociously contagious—even if it was extremely dangerous—then there was time for teams of smallpox vaccinators to react. And, if outbreaks of smallpox were treated as islands in a population rather than as endemic in the populations, mass vaccination would not be necessary. Once doctors had removed the islands in a population, the disease would be gone.

Foege studied smallpox outbreaks in eastern Nigeria and discovered he could draw maps that showed smallpox spreading seasonally from the north to the south each year. Outbreaks decreased in wet months and increased in dryer months, likely due to people dispersing to work their farms in the wet months and coming together in celebrations in the dry months after harvest. In Nigeria, the low-incidence months were September and October. This was significant. April, a high-incidence month, provided 16 percent of a year's total cases, whereas September provided only 4 percent of annual cases. This suggested that vaccinating when incidence of transmission was lowest might break most of the chains of infection, drastically shrinking the potential islands of smallpox.

Foege's analysis was spot on, and later studies would provide ample evidence to back up his conclusions. They would show that all over the world "at any one time a very small percentage of villages (often 1 percent or below) were actually harboring the disease." Smallpox really was a disease of isolated islands in a population. Studies would also show that "vaccination on the day of exposure, or even a day or two afterwards, could still provide protection." This was because the incubation time of the virus in the vaccine was faster than that of the pathogenic smallpox virus. So the race to infect could be won by the vaccine virus, conferring immunity before the deadly virus became full-blown. This also meant that the strategy of attacking an outbreak could sweep away the fuel surrounding the smallpox, even after the first case of smallpox had appeared.

It is easy to think that science only occurs in laboratories or behind desks. And indeed, the eradication of smallpox did reflect some of that. Historians highlight four scientific advances that made eradication of smallpox a real possibility. The first was development of an adequate vaccine. While a smallpox vaccine had been used for more than a century-and-a-half before the eradication effort began, it was not portable or stable in hot climates, where many people needing vaccination lived. Transporting refrigerated vaccine to much of the world was a logistical nightmare. The answer was freeze-dried vaccine, which had been developed in the 1920s. Unfortunately, it was often contaminated by bacteria. Phenol had been added to it, which destroyed the bacteria, but often damaged the virus making up the vaccine as well. Leslie Collier, of the Lister Institute in England, solved this problem by developing a stable unrefrigerated vaccine. He added peptone, a soluble half-digested protein, which prevented damage to the virus. The resulting freeze-dried vaccine could remain stable for years at temperatures of up to 113 degrees F. The dry powder was suspended in a 40 percent solution of liquid glycerin when it was time to be used.

Engineering advances were also necessary in order to deliver the vaccine effectively and efficiently. Before the invention of the jet injector, such as the Ped-o-jet Foege used, the vaccine was scratched into a person's arm using a needle or a rotary lancet. This meant it was not precisely administered, so it often failed to cause an immune reaction. The jet injector Ismach developed provided vaccine take rates of close to 100 percent, and was to be used throughout the Western Africa campaign. In 1961, Benjamin Rubin, a researcher for Wyeth Laboratories, had the ingenious idea of turning a sewing needle upside down and cutting into the eye of the needle, leaving two sharp prongs. The prongs were just far enough apart to hold the exact amount of liquid vaccine needed, due to capillary action. The bifurcated needle, as it became known, also had carefully shaped prongs that prevented it from being stabbed too deeply beneath the skin. Virtually anyone could be trained to pick it up, dip it into a solution of vaccine, then insert the vaccine under the skin of another person. With its close to 100 percent take rates, the bifurcated needle was the perfectly engineered solution for worldwide vaccination in any environment, since it was completely mobile, inexpensive, and could be used with no instructions. It would later be used in much of Asia.

Science can, however, occur outside of a lab as well, and the fourth scientific advance was the development of an effective strategy for vaccinating populations at risk. At its heart, science is based on inductive logic—the examination of detailed data to find general principles. Epidemiologists take vast arrays of information about a disease and its presentation in people, and seek out patterns in that information. Foege was doing just this in the field in rural Nigeria. However, it wasn't as simple as sitting at a desk and going through the data, because he didn't have complete data when he started his analysis. As Foege said, "In retrospect it seems clear—we didn't know how to eradicate smallpox when we started. But this was not a negative. It was a characteristic of all unsolved problems. We are always faced with making sufficient decisions based on insufficient information. If we had waited until all the answers were available, the work on smallpox eradication would never have started. Selecting the target helped develop the appropriate tools and strategy."

One principle of using inductive logic is to always seek out more data. Old data might be wrong or may leave out important variables, so look, learn, iterate...look, learn, iterate. Foege had identified some incorrect assumptions about smallpox, and had also discovered some new variables. He was a powerful, unceasing learner, and he would continue to look for more clues about smallpox as time passed.

While Foege thought his new containment strategy might be more effective than mass vaccination, he understood that just as deductive logic has logical fallacies we are all familiar with, such as contradiction, inductive logic also has its fallacies. The most common one is basing conclusions on anecdotal evidence—on one or two isolated incidents. This is a common fallacy of groups opposing vaccination. Vaccines did not spring up as perfect preventative medicines. Many early vaccines were flawed and caused a small amount of sickness and even death. Even in the United States in the 1960s, the smallpox vaccine killed a handful of people every year. The question a parent must face is: Should their child get vaccinated when there is a one-in-several-million chance of dying from vaccines? Based on the anecdotal evidence of those few occurrences, some parents refuse vaccinations. Others recognize that those odds, while real, must be compared to the chance of sickness and death from the diseases vaccines protect against, which can kill at a substantially higher rate. Nationwide, the few deaths that occur due to vaccines pale in comparison to the number of deaths the vaccines prevent. This is an example of how statistics have come to be so important in science. Statistics allow us to measure risk based on data sets that are imperfect or incomplete, as so many of life's data sets are.

The epidemiologist in Foege understood that his new strategy, which he named "surveillance and containment," had just this logical weakness. No credible scientist would buy into a strategy based on one incident in a small rural province of Nigeria. Foege had to have a lot more proof than one anecdote.

Surveillance and Containment Is Tested

The leaders of the smallpox eradication campaign had already divided Nigeria into regions. Eastern Nigeria was to be coordinated by Foege, Thompson, and Lichfield, who would work with a team of Nigerians from the eastern Ministry of Health. As in every other country, they were supposed to perform mass vaccinations, hopefully covering the entire population. As January came and it became clear they had stopped smallpox cold in Ogoja, Foege could not get it out of his mind how well their very different strategy had worked. He became more and more convinced that mass vaccination was not the correct strategy. Vaccinating the 12 million individuals in their territory would be a huge challenge. There were a multitude of tribes, hundreds of languages, and little infrastructure or communications. Getting 12 million people to line up for a shot was probably impossible, and it seemed quite plausible that the 20 percent they were predicted to miss would be the most likely harborers of the virus. If so, the next year's smallpox caseload might well be the same as the previous year's. That meant lots of boring injecting for little progress.

Foege and his colleagues began discussing their new strategy of surveillance and containment with Dr. Anezanwu, the director of smallpox for eastern Nigeria. Foege says, "We sat down and showed them what had happened, then looked at previous graphs of smallpox in eastern Nigeria where there was a very clear indication that smallpox often came from the north and then swept south. So you had this question in your mind: What if you got on it early and got ahead of it—could you stop it from spreading south each year? We talked about trying a surveillance approach of figuring out where the smallpox was and then concentrating on those areas. It's to the credit of the Ministry of Health people that they were willing to try something new."

There was definitely risk involved for those in the Ministry of Health. If the Americans were wrong, they could simply leave the country. If Anezanwu was wrong, he would have to live with the results. It might ruin both his career and his reputation if his decision set back smallpox control. But Anezanwu was a member of the Ibo tribe, a people proud of their entrepreneurial tradition. He had an open mind, and as he listened he became willing to think outside of the box. The more he learned, the more Anezanwu realized the new strategy might work.

Foege remained confident, but he knew well what was at stake: "In retrospect, you wonder what was going through their minds. Why would they take a risk when everyone else in West Africa was planning to do this one way? Why would they take a risk of doing something that was counterculture?"

Perhaps helping Foege's cause was a rift between eastern Nigeria and the national government, located in western Nigeria. The year before there had been conflicts between those in the west and the Ibo tribe. Every day the newspapers wrote articles about the east breaking away and becoming independent, so that the Ibos, the largest tribe in eastern Nigeria, could rule themselves. The Americans were reassured by politicians that a civil war was unlikely, that there would be some kind of settlement. Besides, even if there were going to be a war, it would take a long time for each side to get soldiers trained and supplied.

Eventually, it was agreed that the eastern Ministry of Health would use Foege's surveillance and containment strategy. For the big eradication push, Foege and his family rented an apartment in Enugu. It was sparsely furnished and their neighbors illegally siphoned off their electricity, but it was home. The plan was to set up a communications network to listen for reports of smallpox cases from healthcare workers, teachers, postal employees, missionaries, and even the public at large. When an outbreak was discovered, the three team members, along with Ministry of Health officers, would race to the site, verify the presence of the disease, and then train the local people how to record cases, vaccinate people, and search out more cases. The three team members played the role of military generals, spreading out maps and drawing circles where they predicted the virus might go, so they could vaccinate people in these places in advance.

Eastern Nigeria stretches from Lake Chad in the northeast, down through scratchy desert-like terrain dotted with round, red mud houses, to a mountainous granite plateau in the middle of the country, and from there on down through the savannah country of Ogoja all the way to rainforest jungles bordering Cameroon in the south. It was vast, and often the three American advisors were gone weekend to weekend. There were countless delays due to broken vehicles, detours around washed-out roads, and requests for medical help for the sick. Foege always took reading materials with him, and at any halt he would find some shade to read under.

During one trip down a dirt road a roadblock suddenly appeared. Their van's driver hit the brakes and nothing happened. He twisted the steering wheel to avoid the blocked road and they veered into the brush, hitting a tree. Everyone was OK. The landowner rushed over, irate that the vehicle had offended his juju—his sacred tree—and demanded compensation. Foege calmly informed him that he had it completely backwards. The tree had damaged his juju, his sacred vehicle, and it was he that was due compensation. Eventually they called it even and the race to the smallpox outbreak continued.

Sometimes a single smallpox case was reported; sometimes there were many. One time Foege saw 1,000 cases in one week. Throughout Africa, villagers sometimes built temporary abodes outside of their village to house people who were sick with smallpox. Past victims were responsible for care and feeding, since it was commonly known that smallpox survivors developed immunity. In one such temporary village Foege saw hundreds at all stages of the disease. The smell of rotting flesh was pungent and sickly sweet. Foege, like the others, had to force himself not to be overwhelmed. Rafe Henderson, an officer working in another area, recalls: "To see a bad case, the reaction is one of horror, of wanting to withdraw. The person is just covered with blood and scabs and the eyes are closed and puffy. Breathing is hard. Your fear is not only that you can do nothing for him, but also your reaction is, 'Oh, my God! Could I get this?'" They had all been vaccinated and had to trust the science behind it or go home. To some, the work proved unbearable, and they did leave.

Some villagers coped in another way. They worshipped smallpox cults, known as Sakpata or Soponna. For them, smallpox was a visit from an unhappy god. Fetishers of the cult might erect shrines, and some priests received the possessions of those who died, as well as practiced variolation—the inoculation of people with the smallpox virus, itself.

It was tough on the three Americans' wives, living in a foreign land, with their husbands racing to Makurdi, then on up the Benue River to Jalingo and across it to Mubi. And race they did, chasing down smallpox cases, training workers, and attending meetings. Meetings where both the national government and the eastern region leaders were present became contentious, sometimes over details as minute as the type of posters used to advertise the campaign. It was obvious that politics were bleeding into the campaign. All supplies came through the national government, and all the workers came from the eastern region's government, so the Americans tried to tread lightly, tried to remain on good terms with both sides. The eastern region became suspicious that the national government was sending in spies, so roadblocks began appearing. Often they were manned by armed teenagers full of alcohol and bravado.

When smallpox hit the state capital Enugu, a decision had to be made. All the trained teams were busy vaccinating in the east. Should they pull teams out to treat Enugu or finish the jobs they were doing? Thompson remembers the argument and Foege's position: "He wanted to pull the teams out of Ogoja to vaccinate Enugu, which I was reluctant to do. But we did just that." Foege insisted on hitting every outbreak. If a cinder escapes from a forest fire, it must be chased down before it can start another big fire.

In Enugu, the three drove around the city, deciding where to vaccinate. As they stopped and laid out the maps, the locals gathered around, suspicious of what the three foreigners were doing. Foege recalls them "planning out vaccination sites on a Saturday for the next Monday. It meant finding a place big enough that you could actually have crowd control and do everything you needed. Suddenly, we were surrounded by police—I was so intent on what I was doing that I didn't even see them pulling up." Foege and his colleagues were arrested and held for five hours before Anezanwu showed up to get them released. It wasn't the only time Foege was arrested. He joked that the evidence proved he was politically neutral, for he was arrested by both sides.

As spring arrived, a civil war seemed a real possibility, so the advisors evacuated their families, returning to do their jobs alone. The roadblocks became more common. Sometimes they were stopped and searched every few miles. They learned that official-looking papers helped expedite their passage, so they began creating their own documents. Many of those who manned the roadblocks were illiterate, so when presented with papers containing apparently official stamps, an officer would often fake literacy to the soldiers who had brought in the foreigners and allow them through. They continued to use a regimen of surveillance and containment, and the vaccinations went forward.

In western Nigeria, the Ministry of Health had barely gotten organized—officials were still preparing for a mass vaccination effort. In comparison, workers in the eastern region were racing all over the country throwing vaccine at every little outbreak. It was analogous to the national government using the 1700s-era British Redcoat strategy of getting everyone lined up, then marching at smallpox as a unit, while the east was using George Washington's tactics of guerilla skirmishes wherever the enemy was encountered. As a result, the national government had barely used any of their supply of vaccine, while the east was quickly doling theirs out. Foege says, "We were going so fast in the east that we were running into criticism from the national government. They were thinking politically about the Ibos and not liking the things the Ibos were doing. We were obviously using vaccine faster than other places because other places were still just getting organized and here we were vaccinating as fast as we could at every outbreak. They said this is actually a national program—we have to keep all the regions at about the same level, and so until the other regions catch up with you, you won't be getting more supplies."

What is a military without bullets or a doctor without medicine? They had been making such rapid progress, and now suddenly they had to quit. When a new outbreak of smallpox sprang up, they had no vaccine. Smallpox would retain its reservoir in the population, grinding out deaths for another year.

A Strange Mission for a Missionary

This was not acceptable to Foege. He had a new vaccination strategy to prove. He knew it was working—if they stopped now, how would they ever prove their method was effective? Foege and a colleague had an idea.

It was 451 kilometers to Lagos and the drive there was fraught with tension. Their intended mission was something Foege never would have imagined doing when he first arrived in Africa two years before. He and his colleague slipped into Lagos and approached the warehouse where the national Ministry of Health's supplies were stored. Inside, they claimed they needed information for a specific part of a jet injector, reconnoitering the warehouse all the while. That night they mapped out a plan. The next morning they returned to the warehouse, and while one of them kept the guard busy, the other slipped back to the vaccine supplies and requisitioned an order, loading their truck full of boxes. The one time missionary had resorted to stealing.

Back on the road, they kept looking in their rearview mirror. A truck had been following them a long way. What would they do if they were stopped? But the truck turned off the road behind them. How long before the vaccine would be found missing? Would they make it back? Their nerves were on edge the whole trip, their necks sore from constantly looking back. Night fell, with many miles still before them, but finally they neared the Niger River and the Onitsha Bridge. The wide river had been muddy the day before when they crossed it. On the other side would be relative safety. There might be consequences to pay—they might be admonished, even deported—but the vaccine would be safely on its way.

Suddenly a roadblock confronted them. Tanks and trucks were lined up in a show of power. Was it the West or the East manning the roadblock? Foege and his colleague pulled up and a guard approached them. He was Ibo, from the East. No one got over the bridge at night, no arms would cross the river. The guard got his boss, who got his boss, and slowly they moved up the chain of command. Finally, they talked to the commanding officer. They explained they had supplies for smallpox vaccination to take into Enugu. The boxes were examined and found not to contain arms. The massive tanks were started, noisy and fuming with diesel exhaust, and rolled back, and the two smallpox warriors drove triumphantly across the bridge, back into the East, carrying vaccine that would soon save lives.

A civil war, known as the Biafran war, started shortly thereafter and changed everything. Foege says, "There were things that made you wonder what you were doing there. I remember thinking, 'Medical school wasn't ever this hard.'" Thousands upon thousands of Ibos in the West were uprooted and forced to move back east. The national government blockaded the East, cutting off all trade. Estimates are that up to a million people starved. Many of the pictures you may have seen of small, starving children with distended stomachs are of eastern Nigerians during the Biafran war. Foege would twice be fired upon with machine guns when he was working in a relief effort during the war.

But the vaccine did get through and made it to the last outbreak of smallpox. There was no communication with the eastern Ministry of Health for a long time, and no one was checking on smallpox cases. Later they did learn, though, that there were no more smallpox cases in the East. The last supply of stolen vaccine had done the job. In only six months' time, using the new strategy of surveillance and containment, Foege and his colleagues had wiped smallpox completely out of eastern Nigeria by vaccinating only 6 percent of the population.

Back at the CDC

Back in the U.S., in Atlanta, Foege had confirmation of his surveillance and containment strategy. He said, "You needed an area large enough so that when the system of surveillance and containment worked, you could actually sell it elsewhere. Eastern Nigeria was that place. If we had not been able to do that, I'm not sure we could have gotten other countries to try this. We had no idea we only had about six months to prove the case before a civil war would break out."

Foege had no trouble selling it to the CDC. "My boss at the time was Don Millar," Foege says. "He was a very energetic guy who just immediately, with no qualifications, thought this was a great idea and we should implement it. And we did implement it in the twenty countries of Western Africa where the CDC was operating. In each country, you have a different rate of take-up, because not only do you have a mass vaccination program already in place, which you can't just drop, because also there was a measles vaccination program along with the smallpox. So you have to continue that program, but now you have another program along the side that is trying to do surveillance and containment.

"I think one of the most successful uptakes was in Sierra Leone where Don Hopkins started a year later than everyone else, but he started right from the beginning with surveillance and containment. Sierra Leone had the highest rates of smallpox in the world at that time, and yet he showed how fast he could bring down smallpox by using surveillance and containment. So while you have some places that were fast on the uptake, others were slower. Nigeria, in the other regions than the east, was somewhat slow in taking this up, and so Nigeria was the last place of the twenty country area to have cases of smallpox. I think their last case was May of 1970."

The program began in September of 1968. In one country after another, within nine months of implementing the surveillance and containment strategy, smallpox disappeared. The success propelled Foege to become the director of the CDC's smallpox unit.

Colossal India

By 1973, the thirty-three countries with endemic smallpox had been cut to four: India, Pakistan, Bangladesh, and Ethiopia. India was the most vexing. It was nothing like eastern Nigeria—the scale of the problem was immense. With a population of 600 million—50 times greater than that of eastern Nigeria—India had over a half million villages and 2,641 cities. A concerted mass vaccination effort was begun in 1962, yet five years later more smallpox was being reported than at any time since 1958. How could mass vaccination work when every day there were another 70,000 infants born?

From Atlanta, Foege wanted to suggest new tactics of surveillance and containment for India, but found himself too far away. "For several years I found myself very frustrated," he says. "We just couldn't seem to get a foothold in India that would work. Some of our best people from Africa went there. They came away so discouraged, telling us that it's never going to work there. I finally went to Dave Sencer, director of the CDC, and I said, 'I don't know if it'll work or not but I do know I can't be criticizing from the sidelines.' So I asked him if he would assign me to India, and he did."

In August of 1973, Foege joined the World Health Organization (WHO) team of Nicole Grasset, a crack Swiss-French virologist, epidemiologist Larry Brilliant, an indomitable American hippie doctor who had moved to India to live in a Hindu monastery, and Zdeno Jezek, a fearless Czech epidemiologist who had arrived after spending five years in Outer Mongolia.

"I lived in New Delhi," Foege says. "The two states that I took primary responsibility for were Uttar Pradesh and Bihar. They had by far the highest concentration of smallpox in India and everywhere in the world. There were lots of people who didn't think that surveillance and containment could work in India because of the high population density and the high level of smallpox transmission. The feeling was you could never get ahead of the transmission because it was so intense."

Still, they planned an active smallpox search for that fall, the seasonal low point of smallpox. It was a daunting project. Four states, cutting a swath across India's northern border with Nepal, provided more than 90 percent of India's smallpox cases. Twenty-two surveillance teams would organize searches city by city, village by village, each team covering 10 million people.

When they started, the reports coming in were overwhelming. Foege says, "When we did our first big search in India in October 1973, I had actually been promoting the idea of a reward for finding smallpox. October was supposed to be a low month for smallpox and so I thought this would be the ideal time to do a reward, when we have the fewest cases that we're going to see during the year. In six days time, our search in four states found ten thousand new cases of smallpox that no one knew existed. If we had given a reward, the bank would have been broken. I am so pleased that the Indians just absolutely refused to consider it."

Foege says, "After that first search, there was a lot of discouragement and some people thought we should stop the searches. I kept arguing that the reason we found so much is because this is the most efficient surveillance system we've ever tried, and the last thing you want to do is stop using it. Let's just wallow and get behind, but let's keep looking, and that's what we did."

Foege ordered sixty more epidemiological officers (a total of 236 from thirty countries were eventually used). It was intense—eighteen hour days, seven days a week. He says, "We were having a monthly meeting in each state that was of high incidence. We were having people come back from the field for a day, reporting on what they were doing, and we were trying to learn as fast as we could and keep making tactical changes every month. We were fine-tuning this just as rapidly as we possibly could. Then, each month, we tried to pretend that we were on top of it and each month it kept getting worse."

The Indians had their best minds working on the problem, too. These included M.I.D. Sharma, a premier epidemiologist who had not taken a vacation in twenty years and had connections all the way to the health minister in the government. Also assisting were C.K. Rao, R.N.Basu, R.R. Arora, Mahendra Singh, and Mahendra Dutta.

Foege says, "Those of us in WHO started traveling by train and finding ourselves in the overnight compartments with our Indian counterparts, and this turned out to be so different than going in for a weekly meeting in someone's office. Pretty soon, a trust level had developed where we actually were discussing the crucial things, even the sensitive things."

There were lots of arguments. Some remember them as personal and vindictive. For Foege and many others though, they were not the dig-in-your-heels type arguments; rather, they were a dialectical debate. Everyone was tossing out ideas, trying to find the best methods that would work. It was iteration toward progress, always moving toward the goal. Listen, learn, iterate.

One of the big lessons they learned was psychological. "To start with," Foege says, "it seems to make sense that the searchers should always have vaccine with them, and when they find a case immediately vaccinate around them. We found out that doesn't work. The reason is that approach biased a searcher to feel they had to do more work if they find a case than if they didn't find a case—you're inducing them not to find a case. You actually have to divide the workload. Searchers only search and report on cases and then containment teams go in to vaccinate.

"In a big city," Foege says, "same as in a country area, we would do a six-day search every month in smallpox endemic areas. In the city it meant going house-by-house looking for cases of smallpox. Cities actually turn out to be harder than villages even though transportation is easier. People are much more anonymous in the city. You might not know whether a neighbor has smallpox. In a village, they will know, but in a city they might not know and they might not tell you even if they do know. So cities are harder; it takes more concentrated work, but it was the same principle of going house-by-house once a month trying to find where smallpox existed at that moment because that's where the virus was at that moment."

Initially, Foege and his team instructed the vaccinators to vaccinate the family of each infected person and twenty households around the outbreak. But that didn't work and the number of cases kept growing. They could barely get to a case before they needed to go on to another.

"I was under great stress and strain," says Basu, an Indian who worked at the health ministry. "Whenever they asked me, 'What is the progress of the program?' I used to tell them the number of cases that month. It was always higher than the previous month. And they used to laugh at me and make sarcastic remarks: 'Dr. Basu says that progress is satisfactory because more of our people are dying.'"

In May of 1974, there were 11,600 new cases in one week in one state, resulting in 4,000 deaths. In the state of Bihar alone, there were 8,664 infected villages. It was the same month India detonated its first atomic bomb, and plenty of pundits noted the irony that they could smash the atom, but could not smash smallpox.

CDC Deputy Director Bill Watson recalls, "It really dismayed people. There was a lot of second guessing. The pressure was on to do it the old way—this new system isn't working."

The surveillance and containment effort came dangerously close to imploding. At one meeting a minister of health insisted on giving up on surveillance and containment and going back to mass vaccination. An Indian doctor stood up and said he had grown up in a poor village: "When there was a fire, the villagers poured water on the burning hut, not on all the houses." The Minister relented and gave them one more month.

They kept trying. When the big numbers came in, they could not even send teams to all the families with smallpox before the next batch of cases came in. But they never quit sending out surveillance teams and never quit sending out vaccinating teams. "It's at least like putting some water on the fire," Foege said, "and decreasing the intensity even if you don't get the fire out." And every case they did stop from spreading provided more and more distance between the outbreaks.

Finally, in June of 1974, there were fewer cases than in the previous month. In July, fewer still. The numbers began reflecting the effects of all of their work. Then the numbers really started dropping. In late 1974, the active search program was expanded, eventually including 33,000 district health personnel and more than 100,000 field workers going house to house, searching for smallpox cases. They began offering monetary rewards for anyone finding a smallpox case. This brought the public into the program, and they reported 11 percent of all the remaining cases. January 1975 only turned up 1,010 smallpox cases in all of India. In February, there were only 212.

"It seemed so sloppy in retrospect," Foege says. "On the other hand, when you see how little we did, and yet we were already influencing transmission, you see how effective that containment can be."

The last case of smallpox in all of India occurred in May of 1975.

"I think the story is so incredible in India," Foege says, "and the thing that turned it around is what you would always like to see in a coalition but don't often see, which is that people suppress their own egos and you get a coalition of people who really work as an absolute team. I don't think the Indian government would have ever have made a major decision without those of us who were now assigned through WHO agreeing, and we from WHO would never have made a decision without the Indians agreeing. It was simply a unit working together."

Bangladesh was the next country to be attacked with surveillance and containment, and then finally Somalia. The last case of naturally occurring smallpox in the world was recorded there on October 26, 1977. Ali Maow Maalin, a 23-year-old man, recovered—but the virus was dead.

Bill and Melinda Gates

After India, Foege returned to the CDC, eventually directing it from 1977 to 1983. He moved on to Emory University in Atlanta as a presidential distinguished professor and also served as executive director at the Carter Center, formed by former President Carter to support humanitarian efforts globally. In 2000, when Bill Gates, the richest man in America, and his wife Melinda were setting up a foundation, Foege was asked to be a consultant. He signed on and was given the task of preparing a reading list for the Gates. Foege says, "I didn't know what the appropriate number of books might be. I sent e-mails to a few trusted friends and asked them what their top five books might be." Foege turned in a reading list of eighty-two books.

Foege says, "I saw Bill a couple of months after that and I asked, 'How are you doing on those books?' And he said, 'Well, I have been so damn busy I have read only nineteen of them.' I didn't know whether to believe him, so I asked, 'Which was your favorite?' He didn't hesitate for a second. 'That 1993 World Bank report was just super,' he told me. 'I read it twice.'" Still skeptical, Foege pressed him on what he had learned from it. Gate's answer proved he wasn't just another rich guy with a momentary interest in doing a good deed. "Not only had he devoured the World Bank report," Foege says, "but on his own he had found the weaknesses. Incredible."

The Bill and Melinda Gates Foundation chose as its goal to "bring innovations in health and learning to the global community." Warren Buffett later joined in, giving substantial sums to the foundation—together they have contributed more than 38 billion dollars. Foege has been an advisor and cannot be more complimentary of the role the Gates are playing in global health: "I think a hundred years from now, when the history of global health is put into some perspective, it's going to be clear that the tipping point was about 2000. It'll be due to Bill & Melinda Gates. They've changed everything. They've made it possible for people to think in terms of better tools, delivery of those tools, and finding the right resources. And they're absolutely serious about the fact that everyone should have the same chance at health no matter where they're born."

Back on the Streets of New Delhi

Other historians might conclude the tipping point occurred three decades earlier. The eradication of smallpox showed what humankind can accomplish when people care enough for their neighbors to use science to help every human on earth, and every human who will ever be born.

There were economic benefits as well. The U.S. was spending 150 million dollars a year to keep smallpox out of the country in the 1970s. Ever since smallpox was eradicated that money can go to other uses. In fact, those savings have virtually paid the U.S.'s contribution to the World Health Organization every year since. Dr. Halfdan Mahler, the director-general of the WHO when smallpox was eradicated, said that the eradication program was a two-billion dollar gift from the poor nations to the rich, because it was the latter that benefited the most financially. With smallpox eradicated, the wealthy nations could pursue global markets with abandon. Amy Pearce estimates that more than 122 million lives have been saved by the eradication of smallpox—lives that also contribute to economies around the world.

How big a deal was the eradication of smallpox? When a young beauty contestant is asked what she wishes for in the world and answers 'peace on earth,' her naiveté is laughed at. Yet many of the same people who laugh sincerely believe that if someone had stopped World War I, World War II, the Vietnam War, the Holocaust, and all genocides, that individual would be celebrated all over the world. Or would they? To put the eradication of smallpox in context, smallpox killed more people in the twentieth century (300 million) than all of the wars, terrorist acts, genocides, and political famines in the entire world combined (188 million, a figure which includes both combat and noncombat deaths). The eradication of smallpox, by measurement, was one of the single greatest achievements by humankind in history. Yet how many know the names of the scientists who contributed the key insights that made it possible?

Most accounts credit the eradication to five scientific advances: cowpox vaccination, pioneered by Edward Jenner; freeze-dried vaccine, perfected by Leslie Collier; the jet injector, engineered by Aaron Ismach; the bifurcated needle, engineered by Benjamin Rubin; and the surveillance and containment vaccination strategy, developed by Bill Foege.

When asked how he feels about his own contribution to the eradication of smallpox, Foege tries to deflect attention to others—D.A.

Henderson, David Sencer, Don Millar, Alex Langmuir—and before long he is outlining that long chain of events that must occur for each dose of vaccine to be put into an individual's arm: hundreds of thousands of health care workers, millions upon millions of teachers and researchers, billions of taxpayers funding it all. He is correct—it is a triumph for all of humanity.

Be sure to press him further, though. Ask him about his own personal satisfaction. Then sit quietly and listen to the poetic image he conveys: "I've been going to India now for 40 years." Speaking of when he first went, he says, "I was very conscious of how many people on the streets had pockmarks." A decade later when he was running all over the country trying to eradicate smallpox, he still saw pockmarks, the telltale markings of those who had survived smallpox. Then ten more years passed and he returned to New Delhi. Vividly, he remembers standing on the street and watching the people pass (India is a country of people; people are everywhere). As he watched children go by—some laughing, others quietly serious—not a single one had a pockmarked face. Then later still, in the 1990s, he returned to New Delhi, and again stood on a street and watched the masses of humanity pass—now not a single person under the age of twenty had the pockmark scars. Just recently he was there again, standing on the streets, looking at the faces, and now no one under the age of thirty had pockmarks. "I think to myself, this is a change that almost everyone walking down the street is unaware of. There is just no memory. Isn't that great? It's fun to see and I still get pleasure out of seeing the tangible results."

Lives Saved: Over 122 Million

Discovery: The eradication of smallpox

Crucial Contributors

Bill Foege:	Developed the strategy of vaccination called surveillance and containment that provided fast removal of smallpox from all populations with as little as 6 percent of the population vaccinated.
Leslie Collier:	Developed a reliable freeze-dried vaccine.
Aaron Ismach:	Developed a jet injector gun that gave reliable vaccinations.
Benjamin Rubin:	Developed the bifurcated needle that could be used anywhere in the world to vaccinate people.
Edward Jenner:	Proved that cowpox virus (vaccinia) could provide immunity to the smallpox virus, giving birth to the concept of vaccination.

Bill Foege examines a patient with smallpox during an epidemic in Eastern Nigeria. Photo courtesy of David M. Thompson, MD, who also worked on the eradication effort.

Henry Gelfand was prodded to look up a tall, clean-cut, practical joker who those at the CDC thought would be perfect as an advisor. Gelfand was skeptical of a "missionary guy," but reluctantly agreed to meet Foege. It didn't take him long to recognize that Foege was highly intelligent, had wonderful people skills, and was full of energy.

"It was necessary, of course, to prevent the virus from traveling out of that protection bubble on contaminated clothes as when a fire crosses a fire line, but the basics for breaking transmission of the virus were remarkably simple and similar to fighting forest fires."
—Bill Foege

The 1967 Biafran War broke out while Foege was in Nigeria. He jokes that the evidence proves he was politically neutral, for he was arrested by both sides. Their intended clandestine mission was something Foege never would have imagined when he had first come to Africa as a missionary two years before.

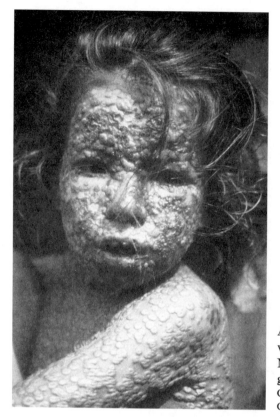

A Bangladeshi girl
with smallpox.
No child will ever
get this disease again.
Centers for Disease
Control / CORBIS.

Rafe Henderson, an officer working in Africa, recalls the dismay: "To see a bad case, the reaction is one of horror, of wanting to withdraw. The person is just covered with blood and scabs and the eyes are closed and puffy. Breathing is hard. Your fear is not only that you can do nothing for him, but also your reaction is, 'Oh, my God! Could I get this?'"

Smallpox is devastating: it destroyed the Incas, the Aztecs, and many American Indian nations, killing as many as 95 percent of their populations.

Abraham Lincoln was coming down with smallpox when he delivered the Gettysburg Address. He wasn't seen for weeks afterward, while he recovered. Both George Washington and Andrew Jackson survived smallpox. Benjamin Franklin's son, Francis, fondly known as Franky, died of smallpox at the age of four.

If the United Nations, through its World Health Organization, never does anything else than eradicate smallpox—it would have to be called a success.

Bll Foege and his wife
in Nigeria in 1966,
dressed to attend the
wedding of a friend.
Photo courtesy of David M.
Thompson, MD, who
also worked on the
eradication effort.

"What is it that is better than science? Better than science is science with heart, science with ethics, science with equity, science with justice."

—Bill Foege

"The philosophy behind science is to discover truth. The philosophy behind medicine is to use that truth for the benefit of your patient. The philosophy behind public health is social justice. That's the important point. Public health programs are attempts at social justice."

—Bill Foege

Bill Foege on Bill Gates: "The guy came to me and said he wanted to learn about public health and he wanted to help. Do you know how many times before I have heard those sorts of things? Rich people say that all the time." The Bill and Melinda Gates Foundation was soon started with Foege as a consultant. With the help of Warren Buffett, it has received donations of more than 38 billion dollars to be used for global health, education, and development. Foege now says, "I think a hundred years from now, when the history of global health is put into some perspective, it's going to be clear that the tipping point was about 2000. It'll be due to Bill and Melinda Gates. They've changed everything."

Chapter 4: David Nalin—
Over 50 Million Lives Saved

ORT: A Revolutionary Therapy for Diarrhea

IT WOULD be hard to find a humbler, more ghastly way to die than by cholera. The typical cholera patient, usually a child, dies amid vomit and excrement, in pain and stupor. Diarrhea—watery and fishy smelling—is the major symptom. The disease is usually found in South Asia, with communities in India, Pakistan, and Bangladesh often the centers of epidemics. Even in modern times it can be a gruesome killer. In the 1980s and 1990s, cholera struck refugee camps in Africa and returned to South America after a 70-year absence. The disease is caused by a bacterium, Vibrio cholerae, which attacks the small intestine. It has the distinction of being one of the fastest killing diseases known. It is possible to have the first symptom of cholera, diarrhea, and be dead three hours later. Usually, however, it takes several awful days for death to come.

Cholera is only one of many causes of diarrhea (from the Greek word for "flowing through"), but, whatever the agent, the symptom can rapidly lead to dehydration and death. In the early 1900s, before chlorination of water and the installation of sanitary sewers, diarrhea was the fourth-leading killer in the United States and the principle cause of death among children. According to the Centers for Disease Control (CDC), there are 1.5 billion episodes of diarrhea worldwide annually, and more than 1.5 million deaths each year caused by this very human indisposition. The standard treatment until recently relied on injecting intravenous (IV) fluid to replace lost bodily fluids and salts, but this was cumbersome, expensive, and rarely available in the tiny villages of Pakistan or the slums of Dacca. Little could be done in these places for the children afflicted by cholera and other diarrheal illnesses. Then, the young American researcher David Nalin and his colleagues developed a simple, effective treatment that could be used anywhere, by ordinary people with no special training.

Their treatment completely changed the landscape of public health in the developing world in a way almost no other discovery has.

The Philadelphia Buddha

David Nalin lives in a modern house tucked away in the hills west of Philadelphia, near the site of the Revolutionary War's Battle of Brandywine Creek. Mostly glass on all sides, the house is not particularly large and most of the space is taken up with a lap-size indoor swimming pool situated on a lower level, which he uses whenever he can. He has a Pakistani chef, so he eats well, countering the effects of all the exercise in the pool. Nalin looks slightly like Agatha Christie's detective Hercule Poirot, slightly round, slightly balding, with a dark small mustache.

Brooding over Nalin as he reminisces is a huge and gorgeous head of Buddha, more than 1,400 years old, carved in a style Nalin describes as Greco-Indian. As gentle-looking as it is large, the bust has a hole in its forehead where an immodestly sized jewel once resided, and features just a hint of a smile, as if this Buddha relishes the wisdom only he has. Nalin explains the style stems from the days after Alexander the Great invaded central Asia, leaving behind Indo-Greek colonies, which is why the Buddha looks a lot like a marriage of classic Greek and Asian sculpture. Nalin knows exactly what he is talking about, for he is a renowned collector of Asian art and an amateur art historian. Traveling throughout the world in the service of science and public health, he filled his house with shelves of brass and copper pots, and statues, and covered his walls with old photographs of Indians, Thais, and Tibetans. One statue in particular reflects the spirit of Nalin. Planted near a window is a figure that once had its hand up to the side of his head in contemplation. Now only the fingertips remains, the hand having been broken off sometime in its history, but the expression remains one of wistful contemplation. The statue is of a bodhisattva, a Buddhist saint who postpones Nirvana to help others through adversity.

Retired, Nalin now has the time to collect and write. Many would say he deserves nothing less. Nalin was instrumental in making death by cholera and other diarrheal diseases unnecessary in much of the world. His is a story of human fallibility, brilliant research, and people who refused to take "no" for an answer. It is of a discovery so big that a UNICEF special report in 1987 said: "No other single medical breakthrough of the twentieth century has had the potential to prevent so many deaths over such a short period of time and at so little cost." It is also one of science's least known major discoveries—few have heard of it. It goes by the acronym ORT.

Treatments

Just hearing the word cholera sends shivers through people all over the world, and every time a natural disaster like a tsunami causes mass flooding, newspapers warn of the threat of cholera due to water contaminated by corpses. Nalin explains that this is a myth—outbreaks usually begin in coastal estuary areas simultaneously with zooplankton blooms, which contain the bacteria. Animals will eat the plankton and humans will eat the animals, or drink contaminated water, and become deathly ill. The bacteria then ride human feces into sewers. If the sewage contaminates water systems, cholera can spread rapidly. For a child, sucking a thumb can be fatal. It is rare for the disease to be transmitted from one person directly to another, as it is not generally contagious.

Essentially, cholera was and is an Asian disease, but sometimes it escapes. Famous victims include former U.S. President James Polk, the composer Tchaikovsky, and both the son and brother of Mary Shelley, the author of *Frankenstein*. In 1854, after repeated epidemics in Europe, John Snow made the link between cholera and contaminated water, and three decades later the German physician Robert Koch tracked the disease with his microscope to the villain—the bacterium V. cholerae. Koch also set up the sanitation rules that are still used today for controlling epidemics. Thanks to Koch's protocol, the sixth pandemic that began in 1899 had little effect in Europe, although cholera remained endemic in Asia.

Before modern medicine, there was no treatment for cholera other than to treat the dehydration. The obvious way to do so, of course, was simply having the patient drink water, but drinking water failed miserably as a treatment. Why it did so would prove a long-term mystery.

In 1831, while Europe was reeling from a cholera pandemic, the Irish doctor W. B. O'Shaughnessy told participants at a medical meeting that the potpourri of treatments being used at the time didn't work. Some doctors used tartar emetics to induce vomiting; others used complex mixtures of herbs, chalk, vegetables, and opium, all to no avail. Castor oil was of no benefit either. Keeping the patient warm with hot blankets and water bottles didn't work. Some even recommended putting a plug in the anus to prevent the flow, which boggles the imagination—the patient may well have exploded. None of these treatments was effective and the death rate for cholera ran as high as 70 percent.

The most popular treatment, used for a host of diseases doctors didn't know how to treat, was bloodletting. If a dehydrated patient's blood was too thick to bleed when lanced, doctors attached leeches to remove the blood. O'Shaughnessy stated that removing blood probably did considerable harm, depriving the patient of even more fluid than was already being lost through diarrhea. By analyzing the blood of a cholera patient, O'Shaughnessy found that the patient had lost not only much water but much of the salt required for life, blood being a saline solution. His recommendation, published in The *Lancet*, was to insert a goose quill needle into a vein and inject a tepid saline solution.

The medical profession largely ignored O'Shaughnessy's suggestion for fifty years, but by the late 1800s, medicine began to lay down a scientific foundation, and intravenous saline solutions proved their worth repeatedly, cutting the death rate of cholera to 40 percent. In the early 1900s, further refinements to the IV solution dropped the death rate to 20 percent. Intravenous therapy worked because fending off dehydration for enough time allowed the body's immune system to rev up and eliminate the infectious agent.

With the arrival of antibiotics in the 1940s, doctors thought they finally had a way to fight the disease itself, rather than the symptoms. Antibiotics did kill the bacteria that cause cholera. Unfortunately, people with cholera can become dehydrated so fast that antibiotics alone can't save the patient. A person can lose up to twenty liters of water a day (10–20 percent of body weight), leading to death by dehydration before an antibiotic can kick in. A vaccine was also tried, but did not prove to be very effective.

In the 1940s, Daniel Darrow of Yale became the leading expert on the use of IV solutions for dehydration. He believed that to treat a patient successfully, one had to know exactly what chemicals were being excreted, and how (and how much) those chemicals changed the patient's body chemistry. He came up with a good approximation of what the body was losing and hence what needed to be replaced. In a series of experiments, he showed that IV replacement solution should include sugar, glucose, salts, potassium and sodium chloride.

Thanks to Darrow, the IV treatment of extreme diarrhea, including cholera, now had the beginnings of a scientific basis. In 1948 Robert A. Phillips, a Navy doctor, successfully used Darrow's formula to cut the mortality rate of cholera to below 5 percent. The problem was that IV therapy required doctors, nurses, equipment, and sterile procedures to prevent infection, and in most of the places where cholera was endemic that was not practicable or even possible. What was needed was a way to rehydrate patients without using modern medical technology. Myriad concoctions—vegetables, fruits, carob flour, bananas, salt water—were devised to rehydrate patients orally in parts of the world without hospitals. But most of what went down came right back up again or was excreted in the flow of diarrhea, so they worked no better than did drinking water. Cholera epidemics in the developing world continued to be devastatingly deadly.

One thing almost every therapy, including IV treatment, did include was a starvation protocol. For cholera, as with other diarrheas, it was believed that the gastrointestinal tract needed rest to recover, so patients were denied food for days.

An Eye Opening, Life-Changing Experience

David Nalin, born April 22, 1941, is a New York native and one of the long line of illustrious graduates of New York City's legendary Bronx School of Science, a highly competitive public magnet school that has produced more Nobel Laureates (seven, all in physics) than most countries. In addition, the school has produced five Pulitzer Prize winners and five winners of the National Medal of Science, and its graduates constitute 1 percent of the members of the National Academy of Science and the National Academy of Engineering.

Nalin began collecting things as a child, starting with baseball cards, insects, and animals. A bathtub was dedicated to keeping his turtles, a cellar to his snakes and salamanders. "I had tolerant parents," he explains. He lived in Manhattan and spent days at museums, occasionally visiting auctions and picking up Egyptian beads or antiquities. A brilliant student, Nalin skipped several grades, and went to Cornell to study zoology at age 16, and then to Albany (NY) Medical College, where he was one of the youngest medical students at the age of 20. On a cross-cultural clerkship he went to Guyana, in South America, and quite suddenly his life changed.

"I applied and participated in my third year," Nalin told his alumni magazine. "Eventually, I would go three times: once when the country was called British Guiana; once during the year of independence, when it was called Guiana; and once the year after independence, when it became Guyana." Having countries change names when Nalin was around would become a trend. It was in Guyana that he learned about practicing medicine in different cultures. "It was an eye-opening, life-changing experience," he recalls.

During the nineteenth century, a Tamil-speaking people from southern India had been brought in as indentured laborers to a coastal section of Guyana. Amerindians, descendants of the original inhabitants, were also present. Nalin was able to put to use the language skills he had acquired at Cornell by quickly becoming familiar with the indigenous languages and developing medical questionnaires needed for research in foreign environments. "I developed one in Guyanese patois and an elementary one in several Guyanese Amerindian languages," Nalin says.

In Guyana, Nalin found the local artisans creating spectacularly beautiful beadwork, and he began collecting seriously. Nalin says his life changed at that time: "Both my cross-cultural medicine and my artistic avocation took root in those years."

"Previously I had a great love for clinical medicine," Nalin explains, "but I found myself being drawn toward research. That surprised me, because I never thought of myself as a researcher. I applied to the National Institute of Health's Department of International Research and

112

was offered my choice of nineteen different laboratories around the world to work in. My experience in Guyana led me to choose a laboratory focusing on tropical, particularly infectious, diseases. Toward that end I developed a protocol for an epidemiological study of blood serum to examine the prevalence of antibodies from viruses transmitted by arthropods in individuals with various cancers, compared to matched healthy controls. This was to have been done at the Middle America Research Unit in Panama. Unexpectedly, the director there turned out to be a Ph.D. of the type who is averse to MDs, and he made it clear that, though he could not criticize the protocol, he did not want to accommodate me under acceptable terms. Just then, Dr. Howard Minners, to whom I reported at the National Institute of Allergy and Infectious Diseases, returned from a trip to East Pakistan [now Bangladesh] and told me that I should consider switching there, as they were doing exciting work on cholera. Initially somewhat aghast (I had just finished brushing up my Spanish for Panama and had no idea what it might be like in East Pakistan), I acquiesced and, of course, never later regretted the switch."

Having only completed the first year of his medical residency, the twenty-six-year-old Nalin found himself signing on to the Pakistan-SEATO Cholera Research Lab in Dacca. The lab was a direct product of the Cold War. In 1959, the United States wanted to further its interests in Asia, primarily to stop the spread of communism. Providing aid was a positive move, so in an agreement with SEATO (the Southeast Asia Treaty Organization) and Pakistan, the United States had opened the cholera lab. It was run by Bob Phillips, an enigmatic figure with a very dark secret.

Nalin did not set out to revolutionize cholera treatment. Like most American doctors, he had been taught little about it, so with the mind of a scientist he set to work as soon as his assignment to Dacca was approved. "It was my habit before undertaking any scientific research project to conduct an exhaustive literature search so as to know the full extent of past work, positive and negative, and to uncover leads for future research buried in the literature," Nalin says.

He learned that the bacteria responsible for cholera are not very robust critters. Indeed, research has found that healthy people rarely get cholera. Researchers have even fed billions of the bacteria to healthy, well-nourished subjects, and none of them became ill. However, if the subjects are given an antacid they immediately become susceptible to cholera. The body's natural defense against cholera infection is stomach acid, which destroys V. Cholerae before it can reach the small intestine to do its deadly work. Most cholera victims have other pre-existing gastrointestinal infections or suffer from malnutrition, which can compromise stomach acid production and render them susceptible to V. Cholerae (in some areas, this describes a third of the population).

No One Believed an Oral Solution Could Work

Before leaving the U.S., Nalin and Richard Cash, another twenty-six-year-old who had also signed on to go to East Pakistan, attended a cholera symposium in California where they heard Dr. Bert Hirschhorn, who had just left the Cholera Research Lab, describe his work. Hirschhorn told them that the current director, the noted cholerologist Phillips, held the hypothesis that cholera poisoned the sodium pump in the human gut, shutting down the transport of sodium through the intestinal wall and making oral therapy impossible. Just the year before, another scientist working in Dacca, David Sachar, had disproved this hypothesis, prompting Hirschhorn and others to speculate on the possibility of an oral solution. The idea was that a solution with glucose might reduce or stop the diarrhea. Hirschhorn performed an experiment that pumped a digestible solution into eight cholera patients' stomachs who were on IV solution. There was a positive result from the study.

In 1962, Phillips had tested oral solutions of water, water with salt, and water with salt and glucose on two patients in exploratory tests in the Philippines. Hirschhorn's trial confirmed Phillips' observation that cholera patients could absorb water, salt, and glucose in a single solution. But because the diarrhea did not stop, and because Phillips had told him of his very dark secret (which he did not dare divulge), Hirschhorn "was not optimistic that ORT [oral rehydration therapy] was feasible as a routine treatment."

Nalin's impression, as he boarded a plane to travel to the other side of the world, was that there was "absolutely no one who believed that an oral therapy could work as a practical thing in rural areas or even in hospitals." Everyone was a skeptic. What formula would you use? How much solution should you give? Physicians doubted there could be a universal formulation, and creating a customized mixture for each patient seemed logistically impossible. Besides, no one believed that cholera victims could drink the amount of fluid that they effuse. How do you get a cholera patient in shock to drink a liter of hydration fluid every hour for twenty-four hours? Additionally, cholera patients often vomit up whatever is in their stomachs.

A Young American in Dacca

Bangladesh, then East Pakistan, is the most densely settled rural nation in the world. It is a country dominated by its geography: The low-lying delta of the Ganges River spreads out as it approaches the Bay of Bengal, covering much of the country. Draining much of northern India and the Himalayas, the Ganges produces the largest delta on earth and a great site for cholera research, for the Bay of Bengal is thought to be a natural reservoir for the bacteria that cause the disease. Cholera has always been endemic there, typically starting when the flooding recedes after the summer monsoon season (which can dump sixty inches of rain on the country in a few months), and whose cyclones cause tidal surges that wash seawater inland.

The Dacca that Nalin arrived in on August 17, 1967, was an old, multiethnic city possessing the architectural ghosts of colonial England, which had ruled the country for almost 200 years before leaving in 1947. Its bazaars were colorful, swarming with people, full of the smells of body odor mixed with smoke, dried fish, and curry. Small taxis honked their horns and jockeyed for position with rickshaws on the crowded narrow streets of the city's historic section.

"My initial impressions of Dacca were wondrous, sometimes bewildering and spellbinding," Nalin says. "On being picked up at the airport by Dr. James Taylor, a beggar boy of perhaps eight years of age grabbed the open window of Jim's Volkswagen as we were about to leave, and with a grimace and tears gestured with cupped hand at his mouth that he wanted money for food. My heart melted, but imagine my amazement when Dr. Taylor spoke a few words to him in Bengali, asking his name and age. The demonstration of interest in him, presented in his native language, fractured his act and transformed him into a smiling, bashful boy who totally forgot about begging. I gave him a coin anyway, and was driven away, mindboggled."

The cholera center was located on the outskirts of the city. Behind it was a waterway that could be navigated all the way to the Buriganga River in old Dacca. Spreading out and forking many times to make up the delta, rivers sometimes widened to more than a mile across, and between them were countless canals. Boats seemed everywhere. There were the ubiquitous *petit patams*—small wooden boats sharply pointed on each end, staffed by one person in the back with a long pole-paddle. Larger ones, called *panshi*, carried cargo such as coconuts or pineapples. In the rainy season the boatmen placed thatched roofs over a small section of the boat arching from gunwale to gunwale. Ferries were another common mode of transportation, and while riding them you might see in the distance a *shampan*—a high-ended, square-sailed boat that looked like a pirate vessel out of the Arabian Sea centuries before. Between all the water, the land was flat. On it farmers grew rice, melons, and jute, a long shiny vegetable fiber that was woven into mats, carpets, chair coverings, and curtains for the thatch-roofed homes of the rural people.

Nalin recalls that "within a few weeks I was taken by boat out on the Buriganga River for the Independence Day boat races. The Provincial Governor, Monem Khan, stood at the head of a yacht, while dozens of wooden launches raced one another, their crews often rowing themselves underwater and capsizing. It was a scene of chaos presided over by officialdom, which proved emblematic of subsequent history."

Another time, he relates, "I was taken to the Buriganga at night to witness the immersion of clay images of the Hindu goddess Durga during Durga Puja, the celebration of the greatest of Hindu mother goddesses in Bengal. As her young worshippers danced around the images, each on a separate boat, they gradually whipped themselves into a frenzy and toppled the images, often eight or nine feet in height, into the river."

David Nalin

No Detectable Pulse

Nalin's first experience with a cholera epidemic wasn't long in coming. In September, cholera broke out in the Chittagong Hill Tracts, along the eastern Burmese border. Chittagong is Bangladesh's largest port city, where the Karnaphuli River empties into the Bay of Bengal. Southeastward, palm trees sway along one of the world's longest beaches, stretching seventy-five miles toward Burma. Today, part of it is a ship graveyard. The world's biggest old ships are run aground at full speed during high tide. When the tide recedes, the huge ships rest incongruously on mud flats, their giant propellers looming eerily out of the water. Underneath, lines of workers trudge out from shore to dismantle the retired ships with blow torches and hammers.

Nalin recalls that when he arrived "patients were dying in their villages because the only hospital was run by Christian missionaries, and the local mullahs had preached that any Muslim who went there would be branded with the sign of the pig. So we had to go out to these remote villages with our intravenous solutions and try to coax parents to let us use them in the huts. A few finally let us do this, and the results were so dramatic that rumors circulated that this could not be cholera after all, because they had never seen a cholera outbreak where anyone survived!" Soon patients began arriving at the hospital in droves.

Even for doctors and scientists, it is hard to be objective in the presence of cholera. There is little that will so concentrate their care as watching a perfectly healthy person turn into a gray, foul cadaver in mere hours. Cholera does its damage in the gut, specifically the small intestine, where the stomach deposits a mixture of partially digested food and water known as chyme. In the small intestine, chyme receives an additional six or seven liters of fluids from ordinary bodily secretions. All this fluid is necessary to maximize contact with the intestinal lining so that nutrients can be absorbed into the bloodstream. By the time the chyme has moved through the twenty feet of small intestine in a healthy individual, 80 percent of the water is reabsorbed and goes back out through the bloodstream to hydrate the body. What's left enters the colon, where most of the rest of the water is reabsorbed, leaving semisolid feces.

It takes about a million V. cholerae bacteria to make someone sick. Those that survive the acidic stomach descend into the small intestine with the chyme. Little propellers pop out of the bacteria that allow them to push their way through the mucus lining the intestinal wall, where they begin producing a poison known as cholera toxin. Phillips had thought that cholera toxin poisons the sodium pumps in the gut. Others thought it destroys intestinal cells. It does neither. Rather it forces the cells of the intestinal wall to expel chloride ions, causing an electrolyte imbalance in the intestine that leads to dehydration.

An ion is an atom or molecule that has lost or gained one or more electrons, making it electrically charged. Electrolytes are simply ions dissolved in water. In our bodies, electrolytes come mainly from dissolved salts such as sodium chloride (table salt), which contains equal amounts of positive sodium ions and negative chloride ions. These electrolytes play vital roles in many cellular processes and our bodies regulate very carefully the concentrations of different ions inside and outside of our cells, using highly specialized ion channels located in our cell membranes. For instance, virtually all our cells have hundreds of thousands, or even millions, of sodium-potassium pumps that are constantly at work exchanging three sodium ions inside the cell for two potassium ions outside the cell. As cells in the intestinal wall pump out sodium ions into the bloodstream, more sodium ions enter the cells from the intestine. Nature has a tendency to create equilibrium, to balance the concentration of electrolytes/ions/salt on either side of a semi-permeable barrier such as a cell membrane.

Cells also have chloride ion channels, and these are what the cholera toxin goes after. When negative chloride ions flood the intestine, sodium can no longer easily move into the intestinal wall cells because of another simple natural tendency—to avoid concentrations of electrical charge. This phenomenon is no different from what happens when you build up electrical charge by scuffing your feet across the floor in winter and then touch a doorknob or other metal object. Nature wants to get rid of the concentration of negatively charged electrons on your body, so the electrons jump from your finger to the metal object and are absorbed by the ground, and you get a shock. Likewise, positive sodium ions are needed in the intestine to balance all the negative chloride ions, and this in turn causes a massive volume of water to travel across the intestinal lining due to a principle called osmosis, by which water moves to balance the concentrations of both types of ions. The chemical imbalance literally sucks water out of the body, the excess liquid cascading into the colon, which can only reabsorb a maximum of about 4.5 liters per day. The only place the rest of the water can go is out of the body.

Diarrhea—watery, profuse, and often painless—begins abruptly, twelve to twenty-four hours after infection. Vomiting may also occur early on. As this vomiting and defecation draws water out of the body, the patient's skin becomes cold and withered, the face becomes drawn, blood pressure falls, and the pulse becomes faint. Death comes from dehydration, after the patient has plunged into shock and coma. In children, the symptoms of dehydration are stark and even quicker: thirst, sunken eyes, dry tongue, shriveled fingertips, and weakness. Pinch the skin on the top of the hands and it remains malformed for minutes. Children are more vulnerable to dehydration because they exchange more than half their extracellular fluids in their gut every day. Adults, in contrast, exchange only one-seventh of their fluids. With cholera, the case fatality rate for children under the age of five can run above 70 percent.

For centuries, every scientist knew that cholera victims could not rehydrate their bodies by drinking water. Nor could they drink salt water, even though scientists knew that salt played a crucial role in the body's ability to move water around by way of osmosis. The key to the mystery of how the body absorbs water in the gut proved to be sugar. Sugar was slow to be recognized as crucial because it plays no role in osmosis, which was the mechanism by which scientists thought the body absorbs water. Although even today scientists don't completely understand all the details, during the 1960s they began to unravel the intricacies of how water is absorbed in the gut. Specialized proteins in cell membranes bind to and transport sodium and glucose (sugar) across cell membranes simultaneously—one glucose molecule for every two sodium ions—and neither can be transported without the other. Hundreds of water molecules are bound to each of these glucose molecules, and this is how water is absorbed. A number of other sugars and amino acids can act in the same way. A driving force behind this transport mechanism is the concentration gradient, or difference, of sodium ions between the intestine and the insides of the intestinal wall cells.

Scientists first learned that water absorption requires salt; scientists then learned that water absorption requires sugar. For water to be absorbed in the gut, all three components—water, salt, and sugar—are absolutely essential.

While this literature was well known to some at Dacca by 1967, it was not at all clear that the absorption mechanisms still worked in cholera-stricken patients. All that the researchers at the cholera lab knew was that Phillips' two-person study and Hirschhorn's eight-person study demonstrated that cholera patients seemed to be able to absorb some salt and water when given a solution of water, salt, and sugar. Doctors don't rely on such small studies in life-and-death scenarios. Perhaps the cholera patients in the studies absorbed the mixture because their sickness was past a critical point. Perhaps they absorbed the mixture because they were over a certain age, or weight. Countless variables could have explained their rehydration and, in any case, they had continued to receive IV treatment during much of the study.

Besides, doctors treating patients with traditional therapy experienced such an exhilarating rush that they were not prone to seek alternative treatments, for not only does cholera begin suddenly and kill quickly, but it can also be conspicuously, even astonishingly cured. As Nalin says, "You receive patients who are about to die, in fact, not infrequently are technically just dead, with no detectable pulse or blood pressure, but with heart and brain still on the edge of irreversible fatality, and within minutes, using standard intravenous fluids matching the ionic composition of their losses, often with added glucose to ward off hypoglycemia, they come back to life. Terminal patients typically received the equivalent of 10 percent of their body weight in intravenous therapy to correct shock, and then they continued to receive intravenous fluids. Tetracycline or other appropriate antibiotic capsules were also given, to shorten the duration of diarrhea to about thirty-two hours on average. This therapeutic miracle stood in sharp contrast to the outcomes of typical emergencies back home in New York: heart attacks, strokes, perforated ulcers and the like, where treatment often failed, never brought an immediate turnaround, and often proved merely palliative or with major ancillary complications."

A Mantra of Failure

As the cholera epidemic rolled on into November and then December, Nalin and his associates worked out of the Memorial Christian Hospital at Malumghat, a portentous name meaning "port of perception." The hospital site had been carved out of the jungle just the year before, and cobras and jackals were still occasionally seen on the grounds. Located up a tidal river from the Bay of Bengal, laborers from the numerous nearby indigenous ethnic groups rafted bamboo down from the jungle, past the hospital, to bayside ports, where it was loaded onto old wooden boats and taken to the offshore islands.

The hospital was now crowded as a result of the cholera outbreak and many patients had family members staying with them. Having only a few dozen beds, the staff erected a large tent next to the hospital to handle the overflow, moving patients there to convalesce. Day after day, Nalin administered IVs. Something of a polyglot, he was learning Bengali so that he could communicate with the local staff and the patients. Particularly poignant were the emaciated children. Sometimes they had to be strapped down to keep them from pulling out their IV tubes.

The center was following up on Hirschhorn's research on a digestible formula, but this time using a new idea. During epidemics there was often a shortage of IV solution because cholera victims need gallons upon gallons of it, all of which requires sterile, distilled water. The idea was that a drinkable solution might function as a supplement to more quickly wean cholera victims off of IVs. This would save IV solution that could be used for other patients.

Rafiqul Islam, a young local physician, had been assigned to work up an oral treatment protocol to be tried in Chittagong. It built upon Hirschhorn's work, and consisted of pumping a liter of sugar and salt solution down a tube into patients' stomachs every hour for eight hours. Nalin and Cash were asked to monitor it. "It was very rapidly a failure," Nalin said. No one died, since all the patients were on IVs and the oral solution could be stopped, but the patients either remained dehydrated or became over-hydrated while they were on the solution.

Early one cold January morning, Nalin sat at a makeshift camp desk in the tent that housed the overflow of convalescent cholera patients. He spent hours poring over and analyzing the data from the failed experiment, thinking about one patient after another. As he thought about each patient's failure, their results echoed through his head: underhydration, overhydration, underhydration, overhydration….

The protocol used a solution that should work; so why had it failed? Underhydration led to dehydration. Overhydration led to fluid flooding the lungs, drowning the patient.

He looked outside of the tent, as the results began to sound like a mantra of failure—underhydration, overhydration, underhydration, overhydration….

Across the tent, he recognized the father of a convalescing infant. He was wearing a sarong-like *lunghi* with a white prayer cap on the top of his head, and he was incongruously feeding his child with a baby bottle. It was the old versus the new. The past versus the future. Killed by cholera, healed of cholera.

"Then, suddenly, it hit me," Nalin says. "Oral therapy had to work, it was the methodology that was the problem. Fluid losses had to be replaced with oral solution volumes that matched or slightly exceeded the volume of losses. I remember a chill going up my spine when I realized this, together with the overwhelming sense of how important this would be to the countless patients who were continuing to be at risk of death in remote, resource-poor affected areas around the globe. And I also sensed that anything that would work in cholera would certainly work in the host of less severe, though often fatal, acute watery non-cholera diarrheas."

Immediately, Nalin jotted down a concept outline for a new protocol. "Having been immersed in cholera treatment using intravenous fluids, I had learned the need to match losses with equal volumes of IV fluids. I also had seen the steady diminution, hour to hour, of cholera diarrhea volume once adjunct therapy with tetracycline or another appropriate antibiotic had been started. For practical reasons, four or six hourly intake and output periods would be used. With diarrhea decreasing with each successive period, it would simply be necessary to give a volume of oral solution to drink which equaled the volume of diarrhea of the preceding intake and output period to ensure positive net gut balance, both matching prior losses and ensuring enough extra to replace insensible [sweat and respiratory] fluid losses and to generate adequate urine flow."

Nalin's brain surged with ideas of how to turn his idea into reality. He conferred with Richard Cash, who immediately agreed to collaborate. These two young doctors, with only three months immersion in this most foreign of cultures and three months familiarity with this disease, were ready to perform groundbreaking science. They were too young to fear failure. They were too new to cholera to feel hopelessly overwhelmed by its heinous power. They were scientists, intent on abstracting an idea from observing data, testing it, and then proving it.

Net Gut Balance

Excited, Nalin and Cash returned to Dacca. The Cholera Research Lab's director, Robert Phillips, was out of town, so they conferred with Kendrick and Ruth Hare, who approved a new trial, Kendrick being the acting director. The trial would have controls so that a comparison could be made between those receiving an oral solution and those receiving only IV solution. Since the patients would be deathly ill, the researchers would use an excess of caution. The trial would take place in the Cholera Research Laboratory Hospital and the researchers would use a specific gravity plasma test for patients receiving the liquid drink, to inform the doctors if dehydration was accelerating. In addition, either Nalin or Cash would always be on call, twenty-four hours a day. There had been centuries of failure in keeping cholera patients normally hydrated with a drink. Would they be the first to succeed?

When the cholera season arrived in Dacca in April of 1968, the researchers chose twenty-nine of the sickest patients at the hospital. All were in shock, as evidenced by having no pulse, and either very low or undetectable blood pressure. They were treated with IV solution until their blood pressure became normal, and then placed into one of three groups:

- Ten were put into a control group which received normal IV treatment.
- Ten were given oral solution by plastic orogastric tube, threaded down their throat into their stomachs.
- Nine were given oral solution to drink.

The oral solution consisted of 4.2 grams sodium chloride, .5 grams potassium chloride, 4 grams bicarbonate of soda (baking soda), and 20 grams of glucose mixed in one liter of water. It was heated to 110 degrees Fahrenheit to aid absorption. The patients in the two oral solution groups started off receiving 750 mililiters of solution (about three cups) every hour if they weighed more than 55 pounds. If they weighed less, they received 500 milliliters. Nalin's breakthrough was that patients should be rehydrated at the rate of their loss of fluid as measured by their output of diarrhea and vomit. Too little rehydration solution resulted in dehydration, whereas too much resulted in fluid overload that could lead to congestive heart failure. They called the difference between what a patient drank and expelled "net gut balance."

Never underestimate the benevolence of special members of the human race. The hospital staff made use of a special bed, the Watten cholera cot, which was a wood-frame cot with a hole in the middle. The patient would lie on the cot, which was covered with a plastic sheet. In the middle of the bed was a hole with a sleeve attached that emptied into a translucent bucket where the flow of diarrhea would collect. Nalin enjoyed working with the staff and recalls that they "were highly skilled and devoted workers who had mastered the simple principles by rote, having had so much experience. They carried out their responsibilities with routine efficiency, skill, devotion, and dignity."

The spirit of cooperation among the staff, which, though mostly Muslim, included several Hindus and Christians, contrasted starkly to the inter-religious fighting that would occur a few years later during the Bangladesh War of Independence. Every four hours, a compassionate staffer would check the calibrated bucket to see how much diarrhea the patient had excreted. Vomit was collected in a basin and measured separately. Then the volume of each patient's oral solution was increased or decreased to match the volume of fluids that he or she was losing.

Nalin says, "As we were taking no chances, we decided that one of us would be with the patients twenty-four hours a day, so we worked in serial eight-hour shifts. The physician on night shift slept in a room connected to our study room, and the nurse on duty would awaken us for periodic intake and output checks and in case of any unusual problem."

Things began as planned, but when Nalin and Cash ended their shifts and the local doctors took over, an unplanned variable intruded on the experiment. Nalin says, "To Richard's and my astonishment, we quickly found out that when our two colleagues were on duty, they would order nurses to restart intravenous fluids on patients who were maintaining themselves well on oral therapy based on no objective indication that there was any need for further intravenous treatment. Apparently they felt that the oral solution, which had not succeeded using their methodology, was not likely to succeed using ours: They simply did not believe in it and thought, apparently, that they were protecting the patients from overconfident researchers."

Dr. Islam, the Bengali physician, recalled, "When we saw a patient, we hesitated just to allow the patients on oral therapy alone. Unless we started IV we feared the patients' deaths. Nalin had a stronger belief. The other local doctors and I had hesitation."

Nalin wanted scientific proof and would have none of such meddling. He aggressively put everyone on notice that the plan to maintain test patients with oral solution alone would continue. To ensure the experiment remained valid, Nalin and Cash decided to alternate twelve-hour waking shifts.

While the staff considered Nalin overconfident in his belief that an oral solution would work, Nalin was confident not so much in his hypothesis, as in the test that would indicate if the oral treatment was not working. The patients' blood plasma was monitored. If the plasma's specific gravity—the ratio of the plasma's density to the density of water—rose above 1.030, meaning that dehydration was serious enough to affect the kidneys, the IV solution was again given until the patient returned to positive net gut balance, meaning he or she was taking in more oral solution than he or she was excreting. Three of the ten patients receiving oral solution from tubes required extra IV solution. Two of the nine who drank the oral solution required extra IV solution.

The results of the trial were spectacular, as Nalin and Cash later reported in their first research paper: "Patients who drank the oral solution tolerated it remarkably well. During the first twenty hours patients drank from 400 to 1050 ml per hour (two to four cups) depending on their requirements as calculated from intake and output records. Of the nine patients who drank the oral solution, three vomited insignificant amounts during the first few hours, but had no vomiting afterwards. Lemon juice was sometimes added for taste. Patients found the solution pleasant to taste."

All twenty-nine recovered from massive diarrhea after about two days. Eventually, all recovered completely. The patients receiving oral therapy, by tube or by drinking a solution, primarily only received IV solution to correct the initial shock and needed 80 percent less IV fluid compared to the controls not receiving the oral solution. The following table summarizes the results of the trial:

Admission Status	Oral Therapy (orogastric tube)	Oral Therapy (Drinking)	Controls
Vital Signs	7 of 10 had no pulse or blood pressure; 3 had B.P. of 70-80 mm systolic	5 patients had no pulse or B.P.; 2 had B.P. of 80 mm systolic; 1 had B.P. of 110 mm; 1 not recorded	All patients had undetectable pulse and blood pressure on admission.
Male/Female	7/3	5/4	6/4
Average age	26 (17 to 46)	24 (10 to 40)	29 (11 to 60)
History of vomiting	9/10	8/9	9/10
Gross stooling rate pr hr: First 4 hours First 24 hours Total study period	.50 liter .46 liter .35 liter	.55 liter .50 liter .30 liter	.46 liter .33 liter .24 liter
Body weight: Beginning of study End of study	93.7 lbs 95.2 lbs	80.4 lbs 82.4 lbs	88.2 lbs 90.0 lbs
Hourly vol. of oral therapy: First 24 hours Total study period	571 ml 513 ml	618 ml 462 ml	
Hourly increase in net positive gut balance: First 24 hours Total study period	98 ml 100 ml	125 ml 108 ml	
Stool culture for V. cholerae	Last positive culture 2nd hospital day in 80% of cases; all negative by 4th day	Last positive culture 2nd hospital day in 80% of cases; all negative by 4th hospital day	Last positive culture 2nd hospital day in 70% of cases; 90% negative by 5th hospital day
Avg. duration of diarrhea before admission	13 hours	12.5 hours	11 hours
Avg. duration of diarrhea per hour after admission	39 hours (+-10)	51 hours (+-7)	44 hours (+-11)
Total IV vol. per patient	4.3 liters	3.6 liters	19.3 liters

Comparison of Patients Receiving Oral Therapy by Orgastic Tube, Patients Who Drank the Oral Solution, and Controls.

Nalin and Cash quickly wrote up the results and published them just four months later in *Lancet* with the title "Oral Maintenance Therapy for Cholera in Adults." Even though Phillips was out of the country, they included his name on the paper because Nalin wanted to recognize "his pioneering glucose work."

The paper concluded: "Our findings indicate that an oral solution containing glucose and electrolytes can eliminate the need for over three-quarters of the intravenous-fluid requirements in the therapy of acute cholera in adults. The ingredients of the oral solution are cheap and widely available in virtually all areas affected by cholera. The solution need not be sterile and it can be made up with any suitable drinking water. Ingredients could be pre-weighed and stockpiled for use in cholera epidemics.

"The drastic reduction in the need for intravenous fluids which results from the use of an oral therapeutic solution should make it possible for cholera treatment centers with limited supplies of intravenous fluids to reduce the mortality from cholera to a level previously not possible in the absence of abundant intravenous fluids. Mild cases of cholera (without shock) may be treated with oral solution alone."

Now the acronym becomes obvious. ORT stands for oral rehydration therapy—a drink that cures the major symptom of diarrhea—dehydration. Their experiment had not been a full-fledged test of oral rehydration treatment, for all the patients had first been started on IVs until their blood pressure was under control. But they had taken the first step in proving that drinking an oral solution could work and could provide a measurable therapeutic advantage.

Phillips' Dark Secret

Aftger they had written up the *Lancet* paper, Nalin and Cash began planning further studies. The Cholera Research Laboratory had a field hospital at Matlab, a remote outpost forty miles east of Dacca that was the center of annual epidemics. It would be an ideal place for a trial since it reflected the realities of health care throughout the developing world, where diarrhea victims lived without access to modern hospitals. The trial needed to be large enough to prove the concept definitively to the scientific community, so it had to include hundreds of patients. The researchers decided they would throw away the tubes for feeding the solution into patients' stomachs and go after the holy grail: a drinkable solution not preceded by an IV solution. They also wanted to go after an even more general result. There were always patients who showed up with diarrhea not caused by cholera. Nalin was optimistic that a drinkable solution could rehydrate those patients as well. In addition, they could test if adding the amino acid glycine to the solution would improve the outcome, following up on prior research done by physiologists showing that adding glycine to glucose increased intestinal sodium absorption.

Phillips was in the U.S., so Nalin and Cash sent their plans to John Seal, the National Institute of Health administrator overseeing the Dacca lab from the U.S., fully expecting cooperation and praise. But the two received a devastating response. In a letter they suspected Phillips had influenced, Seal criticized the ideas behind the new trial and suggested it be delayed and reviewed by the Technical and Clinical Research Committees. The prolonged delay indicated in Seal's letter would likely spell the death knell of the trial, since the cholera season would be over by the time of the committee meetings the next year. By then Nalin and Cash, the driving forces, would also likely be back in the States, their tours over.

Nalin and Cash were dumbfounded. Their last trial had been such a success. They were not moving too fast; they were planning on taking the necessary baby steps that medical advances require. Plenty of safeguards were built into the experiments to ensure the patients' safety. Plus, there was likely to be a shortage of IV supplies during the next epidemic, so their trial might save lives that otherwise would be lost. What could be the problem? Even the staff had expressed unbridled enthusiasm for their study. So why was Phillips so intractable?

Even though Phillips had just the previous year won a Lasker Award—the most prestigious American medical award—for his work on cholera, he was not a happy man. At times he showed signs of alcoholism and depression. When pressed, Phillips retorted that he did not want his clinical researchers working on practical field problems. But that hardly seemed to ring true. Successful clinical research logically leads to practical field trials.

What Nalin and Cash did not know was the secret Phillips had told Hirschhorn. When Phillips was doing his exploratory research in the

Philippines in 1962, he thought he had found a magic bullet that would stop cholera diarrhea. Phillips, a Navy captain who ran the Navy's medical research unit in Taipei, Taiwan, had been working with cholera treatment for fifteen years and had reduced his patients' mortality rate to below 1 percent using the latest IV electrolyte protocol. When his exploratory work had shown that a water/electrolyte/glucose solution could be absorbed, he had formed his hypothesis that the sodium pump in the intestine was poisoned by cholera and that his "cholera cocktail" undid the damage and stopped the diarrhea. He set up a larger experiment that would administer the oral solution at the same time patients received IV solution. Of the thirty patients in the trial, five died. The oral mixture, combined with IV solutions, caused too much water to be retained in patients' bodies, creating an electrolyte disturbance with fluid overload that led, according to one observer, to congestive heart failure.

Phillips was heartbroken. His normal IV protocol had reduced cholera mortality to below 1 percent, but with his magic bullet therapy mortality had skyrocketed to 17 percent, resulting in five deaths that would likely not have occurred otherwise. Devastated, he dropped most of his research on cholera treatment and began a slow descent into guilt. Worse yet, he did not publish the results concerning the deaths.

Phillips did not want another trial, even with Nalin's improved protocol. Completely ignorant of Phillips' fateful Philippine tragedy (Nalin did not learn about it until 1971, when he was back in the U.S.), Nalin and Cash pressed on until a telegram arrived directly from the NIH, blocking the trial. That left them in a quandary. They were forbidden from performing the trial, yet they firmly believed they had found a therapy that could be taken into the field and would save lives. Their options were limited. If they went ahead with the study on their own, the repercussions could be dire. Technically they were Public Health Service officers under the realm of the military, which meant they could face court martial. Yet if they did nothing, no lives would be saved and no scientific advance would occur. The scientific community would not believe their discovery based on a trial with only nineteen patients. Only a large-scale trial would move their advance forward.

Finally, they realized the better tactic would be to try to end-run Phillips. They turned to Henry Mosley, head of epidemiology at the lab, for advice. Mosley technically worked for the Centers for Disease Control (CDC), not the NIH. He came up with an elegant solution. "Simple," he said. "I'll send you four CDC officers. You tell them what to do. You won't actually be doing the test and violating orders." Mosley contacted Alex Langmuir of the Epidemiological Intelligence Service. Langmuir recognized the revolutionary nature of the discovery and agreed to the ruse. They all knew that Phillips was not interested in the CDC unit and was unlikely to review all the proposed studies.

ORT: Born at Matlab Bazaar

In October of 1968, cholera hit Matlab, as it did every fall. Nalin, Cash and others from the cholera center traveled there by speedboat, a five hour journey.

Nalin remembers Matlab Bazaar as a very primitive outpost: "There was a single-story brick and cement building of several rooms for the hospital, on a slip of land which had been picked because of its history as the center of annual cholera outbreaks. The hospital was connected to the bazaar itself by a rickety bamboo and slat wood bridge over a canal, and boat people, a kind of riverine nomadic Bengali gypsy clan, often moored their boat homes on the bank. The hospital was run by one doctor, the late Mizanur Rahman, who had trained his helpers and nurses to treat the patients and run the place. We visiting investigator Young Turks were housed on a floating barge inherited by the hospital, whose barred windows were a reminder of its previous incarnation as a jail boat under the British Raj. It was inhabited by some of the largest roaches I had ever seen. One had to be careful during meals, as the local crows were expert at suddenly swooping down on unsuspecting diners and making off with the food. Ambulance boats would come in and go out over the day, ferrying cholera patients to and from the hospital."

Nalin was by now conversant in Bengali and had designed a questionnaire to interview patients to obtain their medical histories. He says, "The hospital environment was a bit hectic, especially when epidemics were at full blast with hundreds of patients admitted, many in shock, every day. There was a scale at the entrance where patients could be quickly weighed (held by an attendant if they couldn't stand, then deducting his weight from the total). In an instant, an IV drip was started, infusing over an hour or two the equivalent of 10 percent of the patient's body weight to bring back a pulse and blood pressure and correct shock."

Once the IV was started, patients were tested for cholera, and then given from 750 to 1500 ml per hour of oral rehydration solution to drink for the first four hours. After that, the amount to drink was adjusted to the amount they expelled from their bodies. When a patient came into positive net gut balance, meaning more fluids were being taken in than expelled, he or she was removed from the IV solution. If the patient then went negative "net gut balance," he or she was returned to the IV solution.

The protocol worked quickly. The result was that 78 percent of the patients were in positive net gut balance within four hours, and 87 percent within eight hours. Once net gut balance was achieved, a return to IV was necessary for only four out of 135 patients. As weeks went by and more and more patients steadily improved after drinking the solution, the doctors began to treat the less serious patients with the drinking solution alone, without ever starting an IV. When this had been shown to work beyond any doubt, they administered the drinking solution to five severely ill patients without first administering an IV. The blood pressure of even these gravely ill patientis returned to normal, and they never required treatment by IV infusion.

The comparison group for the field trial was the previous year's group of patients. Only 30 percent of the amount of IV solution used the previous year was needed for a similar number of patients. When Nalin and Cash wrote the paper describing the trial, their conclusions were revolutionary:

"After the initial trial was completed, the combined intravenous-oral therapy regimen became the routine method of treatment at the field hospital. All outpatients were successfully treated with oral therapy alone and an additional 445 inpatients, 350 of whom had cholera, were successfully treated with oral-maintenance therapy.

"A total of 580 patients were treated with oral therapy during this epidemic, with an estimated saving of 3,250 liters of intravenous fluids....

Medical and paramedical personnel were easily trained in the preparation and use of oral solutions. Oral therapy practically eliminated the need for intravenous fluids in patients with mild cholera. It seems possible that if patients with severe cholera were treated from the time of onset, they might be maintained on oral therapy alone. The public health implications of oral therapy are obvious; it is inexpensive, simple, and effective. Its widespread use should reduce mortality due to cholera even in remote districts where little or no intravenous fluid is available."

The very end of the paper included a distinctive acknowledgement: "The Authors would like to express their gratitude to Dr. W. H. Mosley...for his invaluable assistance in carrying out this study."

Further Discoveries

In biology, a big breakthrough experiment often inspires numerous additional investigations. In a follow-up to their landmark study, Nalin and Cash showed that ORT can treat even the most severe cases of cholera. They then proved that it works on children. In fact, later studies demonstrated that infants as young as one month old can be given ORT. The addition of glycine was also a success, substantially reducing both the duration and volume of diarrhea in cholera patients. This discovery set the stage for numerous other studies refining and improving the solution. Nalin and Cash also demonstrated that cholera patients could be fed food along with the oral solution, not long after shock was corrected, overturning the long tradition of starving patients for several days. In fact, nourishment can help a patient fight off the cause of the diarrhea.

A big surprise—which perhaps shouldn't have been a surprise at all—was discovered by Hirschhorn when he returned to the subject years later in a study of Apache children in Arizona. His studies revealed that people instinctively know how much solution to drink to properly rehydrate their bodies. It is in a doctor's nature to want to prescribe specific doses of treatments. But it turns out that without a doctor's prescription, even children often drink as much as they need and then stop, although certain conditions may inhibit this ability. In hindsight, this makes a lot of sense: People live their whole healthy lives regulating the amount of water in their bodies, and being sick doesn't impair their ability to do so.

Most significantly, and surprisingly, to many doctors, Nalin and Cash demonstrated that ORT is "as effective in the non-cholera diarrhea patients as in cholera patients." Those patients with diarrhea whose tests came back negative for cholera were rehydrated by ORT, just as those with cholera were.

The importance of this finding was monumental. While cholera raises our awareness of diarrheal dehydration because it comes in attention-grabbing epidemics, it accounts for at most 10 percent of the cases of diarrhea sickness in the world. Diarrhea is also a major symptom of dysentery, caused by bacteria or amoebas; traveler's diarrhea, usually caused by E. coli strains; and norovirus, which contaminates food and water (in 2002, there were norovirus outbreaks on twenty-five cruise ships, sickening 2,648 passengers).

The most prevalent cause of diarrhea is the rotavirus, which, because poor sanitation is not its vector, affects all income groups. Spread by direct human-to-human contact, the CDC estimates that rotavirus causes 39 percent of all childhood hospitalizations for diarrhea worldwide and up to 50 percent of all deaths. ORT has proven to be the main line of defense against dehydration from diarrhea, no matter the cause. In fact, in 1972, Hirschhorn concluded that IV use in the treatment of diarrhea was "old-fashioned" and said that ORT was clearly the superior treatment.

A Pinch of Salt with a Fistful of Gur

"To save the life of a person with diarrhea is probably the cheapest health intervention you can think of," said David Sack, a past executive director of the International Centre for Diarrhea Disease Research in Bangladesh. Indeed, ORT packets could be produced for eight cents apiece.

But this knowledge had to be conveyed to people in many parts of the world who had few educational resources and little access to media. To spur the educational effort, Nalin traveled the globe as a consultant for the World Health Organization (WHO), starting national ORT programs in Costa Rica, Jamaica, Jordan, and Pakistan. The remarkable thing about ORT is that it does not have to be administered by the medical community—anyone can use it, provided he or she has the necessary knowledge.

"We realized for this to have optimal effects, we really had to get it out of the hands of doctors and nurses and into the hands of experienced mothers," Nalin said. "One tool that proved very successful was to teach doctors and nurses that they had to communicate three or four rules to a mother when the mother is concerned her child is sick. Are the child's eyes sunken? A mother knows better than a doctor when a child is ill. At the first signs of diarrhea, take out a packet, mix it with a liter of water, and start giving it to the child every half-hour or so until the child looks normal and starts to pee. We told her to watch the child's eyes; give fluids until the eyes return to normal. We taught her to keep pinching the skin on the back of the hand—it will 'tent' if the child is still dehydrated and needs more oral solution. Mothers were sent back to villages with the packets. That sometimes met resistance from the local medical community, which had been making money giving IVs or charging several dollars for packets that cost pennies. But one mother back in the village would spread the word."

In 1971, less than 1 percent of children in the developing world had access to ORT. By 1980, access had increased to 30 percent and by 1990 it was at 60 percent. Today, total production exceeds 800 million packets a year.

Nalin's regimen completely changed how doctors treat diarrhea. Now doctors recommend drinking fluids as soon as diarrhea begins. They also realize that prolonged fasting can impair the function of the intestines, so they encourage solid food nourishment as well. And if diarrhea persists to the point of dehydration—usually first evident by lack of urine—ORT is the first line of treatment. Once the body can maintain hydration and nourishment, the immune system will fight off most causes of diarrhea in a few days and the patient's health will return.

After ORT was introduced widely in the 1980s, the fatality rate for infants under the age of one with diarrhea from any cause fell by 65 percent compared to the pre-ORT era. The fatality rate for children with diarrhea between the ages of one and five dropped 62 percent. The annual number of child deaths around the world attributable to diarrhea fell from over 4.5 million a year in 1980 to about 1.5 million today. In a 2007 article, the *British Journal of Medicine* estimated that ORT has saved the lives of more than 50 million people. ORT, of course, only affects the outcome of diarrhea, not the cause. The number of diarrhea cases has not dropped precipitously. Since ORT only treats the symptoms, prevention is the best method of avoiding diarrhea altogether. Clean water, better sanitation, breast feeding, vitamin A supplementation to aid the immune system, and rotavirus vaccination are all important preventative measures necessary to lessen the incidence of this virtually ubiquitous human infirmity.

In 1993, post offices in Bangladesh began stamping an odd phrase on envelopes. After the Bangladesh War of Independence in 1971, Abed Fazle Hasan, who had been an accountant working for Shell Oil in Chittagong, wanted to rebuild his country. He formed a private foundation, the Bangladesh Rural Advancement Committee (BRAC), and quickly realized the promise of ORT. There was no industry producing packets of solution at the time, so Hasan, working in his own kitchen, formulated oral rehydration solutions and then tested them in laboratories. He found that the proper home mix for ORT was three pinches of salt and two finger scoops of sugar.

BRAC, with Cash as its major consultant, promoted this recipe by sending women trainers into rural Bangladesh. However, the trainers found that some mothers remembered the formula in reverse order, making it unpalatable. Their recommended solution was to keep the pinch of salt, but change the sugar to a fistful. Sure enough, it tested well in the laboratory. BRAC sent 10,000 female health workers into the field, essentially turning a scientific formulation into a home remedy. They taught ORT to 13 million illiterate mothers, and to countless children in classrooms that BRAC set up. It became such a culturally popular program that it led to the aforementioned stamp:

Mix with much care,
Good water, a liter,
A pinch of salt with a fistful of gur,
Remove the menace for good

ORT acted as a catalyst for Abed Fazle Hasan's efforts to transform his society, his goal being to empower those in poverty. BRAC has since promoted gender empowerment of women, microloans to lift people out of poverty, and education (there are 47,602 BRAC schools in Bangladesh). Now his model is being implemented internationally.

In the U.S., ORT Never Gets Past Gatorade

Even people of science do not always practice science: Most of the U.S. medical establishment still uses IVs to treat dehydration from diarrhea. As Joshua Ruxin, a historian who has extensively researched ORT, says, "A superior therapy, even when supported by an ironclad physiological paradigm, may not necessarily be employed. The formidable and persistent ignorance of the Western medical establishment, which continues over 25 years after the discovery of ORT, is phenomenal." Not surprisingly, without the encouragement of doctors most American parents don't even know that drugstores sell ORT in packets or solution over the counter.

ORT is an example of a less technological solution being superior to a more complex solution, making some parents, who believe that the best treatment for their child is what costs the most money, reluctant to use it. More cynical critics suggest that the medical industry is simply greedy. The cost of ORT is a few dollars a treatment. The cost of putting a child or elderly person on an IV, and often keeping him or her overnight in a hospital for observation, can run thousands of dollars. Estimates suggest that ORT use could save billions of dollars annually in the U.S.

It's especially odd that ORT hasn't caught on as a treatment for diarrhea in the U.S., considering that most Americans have used an oral rehydration solution. In the South in the 1960s, football players were falling from heat exhaustion like flies on DDT. University of Florida assistant coach Dewayne Douglas, himself a past player who had experienced heavy sweating and no urine output during games, asked physicians at the college to develop a drink for the football team. They discovered the water-salt-sugar coupling research and formulated a solution for the players to drink. The rest is history. That year the moribund Florida Gators football team began using Gatorade and finished 7–4, winning many games in the second half when fatigue takes its heaviest toll on players. It was their first winning season in more than a decade, and the next season they went 9–2. Soon, every team was using Gatorade, which is simply a scientifically formulated oral-rehydration therapy.

While Gatorade works based on the same physiological mechanisms as ORT, it is not a substitute for ORT in diarrhea cases. Gatorade is formulated for healthy athletes, chiefly to replace sweat loss, and not for sick children or adults who have diarrhea, which requires a different solution. In order to make it easier for busy Americans to take, Abbot Laboratories created a ready-mixed ORT drink for diarrhea, available in a bottle, called Pedialyte. And only in America does taking medicine have to be fun. To encourage its use, Abbot began producing Pedialyte Freezer Pops in numerous flavors. So maybe ORT will yet catch on in the United States.

Life Is a Candle

Back at his house, sitting in the shadow of the giant, smiling Buddha, it is obvious that Nalin has retained his childhood collecting proclivity. Among other things, he has collected two highly prestigious awards. Abe and Irene Pollin, best known as the owners of the Washington Wizards professional basketball team, offer an esteemed medical award every year. Nalin, along with three other scientists instrumental in the development of ORT—Norbert Hirschhorn, Dilip Mahalanabis, and Nathaniel Pierce—received the 2002 Pollin Prize for Pediatric Research. In 2006, Nalin, Mahalanabis, and Richard Cash received the Prince Mahidol Award in Thailand for their life-saving advance. The Nobel Prize should follow.

Through the years, David Nalin collected art wherever he went. Although he does not consider himself religious or spiritual, he was attracted to personal items of worship more than grandiose objects. He says, "I often found myself drawn back to the immediacy and ingenuous charm and humor of devotional folk images, and to the wholly satisfying formal integrity of the vessels used in their worship: pots to be filled with holy Gangetic water or with milk offerings for serpents…smoking camphor tablets swung by dancers before the goddess during puja…brass platters with mirror-like polish for offerings of Bengali sweets. Also incense holders—a cornucopia of vibrant forms—and simple bowls which, when struck, hum with a resonance evoking another era and another reality."

Nalin and his brother, also a collector, have given away more than 1,000 pieces of art to museums. He has also begun giving today's students the same opportunity to have first hand exposure to health-care practices around the world, in the hope that they will learn some of the same lessons that Nalin himself learned in Guyana and Bangladesh. In 2006, he pledged $200,000 to establish the David R. Nalin 1965 Endowed Fund for International Research. It is to be used to send two students abroad each year for nonsectarian medical research. Nalin says, "The exposure to Guyana whetted my appetite for international research. It changed my life. I hope this gift will change the lives of others."

Some people collect Buddha quotes like David Nalin collects art; many could be placed alongside his life. Here's one: "Thousands of candles can be lighted from a single candle, and the life of the candle will not be shortened. Happiness never decreases by being shared."

But here is perhaps a more compelling quote relating to the story of ORT: "There are only two mistakes one can make along the road to truth: not starting and not going all the way."

Robert Phillips began the search for a magic bullet. David Nalin went all the way.

Lives Saved: Over 50 Million

Discovery: Oral Rehydration Therapy (ORT)

Crucial Contributors

David Nalin: Had the key insight that ORT would work if the volume of solution patients drank matched the volume of their fluid losses, and that this would drastically reduce or completely replace the only current treatment for cholera, intravenous therapy. Led the trials that first demonstrated ORT works, both in cholera patients, and more significantly, also in other dehydrating diarrhea illnesses.

Richard Cash: Provided crucial collaboration, working with Nalin on the key experiments.

Robert Phillips: Made the first attempt at ORT, demonstrating that a glucose solution might be absorbed in cholera patients' guts.

David Sachar: Confirmed Phillips' observation and proved that coupled glucose/sodium/water absorption were intact in cholera patients.

Norbert Hirschhorn: Confirmed and clinically expanded Phillips' observation that coupled glucose/sodium/ water absorption were intact in cholera patients.

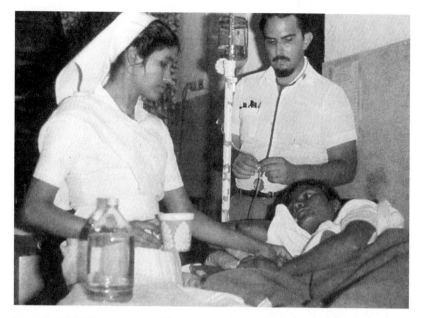

Dr. David Nalin turning off an IV so the cholera patient can begin drinking the ORT soution (at a trial in Dhaka, Bangladesh). Photo courtesy of Dr. David Nalin.

It takes about a million V. cholerae bacteria to make someone sick.

While ORT was proven to work on cholera patients, David Nalin and Richard Cash soon proved it worked to rehydrate patients with most diarrhea diseases.

"No other single medical breakthrough of the 20th century has had the potential to prevent so many deaths over such a short period of time and at so little cost."
—G.A.Williams, UNICEF special report: *A Simple Solution, 1987.*

After ORT was introduced widely in the 1980s, the fatality rate for children with diarrhea fell by over 60 percent. The number of child deaths worldwide from diarrhea fell from over 4.5 million a year to about 1.5 million today.

Today, total production of ORT packets exceeds 800 million a year.

Gatorade is based on the same physiological mechanisms as ORT, although it is not a substitute in diarrhea cases because it is formulated for healthy athletes, chiefly to replace sweat loss.

A recovered cholera patient, surrounded by his record-breaking
number of one-liter bottles of IV fluid that had been used
to replace his fluid losses. ORT replaced most IV treatment.
Photo courtesy of Dr. David Nalin.

A person with cholera can lose up to 20 liters a day of water, 10–20
percent of body weight, leading to death by dehydration.

Cholera has the distinction of being one of the fastest killing diseases known. It is actually possible to have the first symptom of cholera—
diarrhea—and be dead three hours later.

People fear cholera due to water contaminated by corpses. This is a myth.
Cholera corresponds to zooplankton blooms, which contain the bacteria.
However, water contaminated by the feces of people suffering from cholera
can transmit the disease.

Researchers have fed billions of the bacteria that cause cholera to healthy,
well-nourished subjects, and none of them became ill. This is because cholera is killed by stomach acid. But if the subjects are given an antacid, they
immediately become susceptible. Most cholera victims have gastrointestinal
infections or malnutrition, which can compromise stomach acid production,
rendering a person susceptible to cholera for life. In some areas, that describes a third of the population.

Robert Frost's son Elliot died of cholera at the age of three.

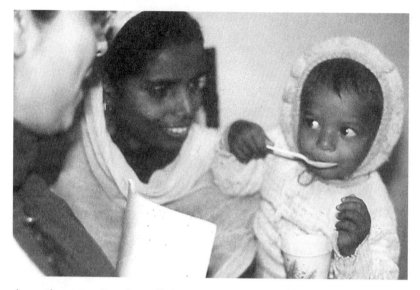

A mother at a diarrhea clinic, giving her child ORT. Photo courtesy of Dr. David Nalin.

Infants as young as one-month-old can be given ORT. Children are more vulnerable to dehydration because they exchange more than half the extracellular fluids in their gut every day. Adults, in contrast, exchange only one-seventh of their fluids. Untreated, the fatality rate for children under the age of five with cholera can run above 70 percent.

In the early 1900s, before chlorination of water and sanitary sewers, diarrhea was the fourth-leading killer in the United States, and the principle cause of death among children.

Diarrhea—from the Greek word for "flowing through."

The Watten cholera cot. The patient lies on the cot, which is covered with a plastic sheet. In the middle of the bed is a hole with a sleeve attached that empties into a translucent bucket, where the flow of diarrhea collects and can be measured. Photo courtesy of Dr. David Nalin.

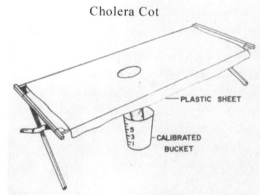

Cholera Cot

PLASTIC SHEET

CALIBRATED BUCKET

Chapter Five: Norman Borlaug— Over 245 Million Lives Saved

A Green Revolution to Enhance Nutrition

THE TALL, scrawny man in the checkered shirt, brown pants, and work boots sat under a broiling Mexican sun, on a small, folding camp stool in the midst of an enormous field of ripening wheat. The dust from the fields and nearby road coated his face, while a handkerchief wrapped around his forehead failed to keep the sweat out of his eyes. In one large, weather-browned hand he held a pair of needle-nosed tweezers. The other hand gently encircled a delicate head of wheat, a tiny fleck of white or yellow hinting at the grain that would eventually emerge. With meticulous precision, his hands as steady as a surgeon's, he used the tweezers to probe the barely formed flower and pluck out each tiny stamen (the male part of the plant), being careful not to disturb the plant's ovary, or female part. Then he slid a small glassine envelope over the wheat head, folding over the top and fastening it with a paper clip. In five days, he would return to this same plant, remove the paper clip and slip in the stamen of another type of wheat in the hope that the two parts would create a third: a new breed of wheat capable of feeding the world.

Finished with the first plant, he moved his camp stool over a bit and started on the second. Then the third. Then the fourth. Day after day, from sunrise to sunset, the man worked, never looking up, his muscles growing tense with fatigue, his eyes red with grit. As night fell, he spread a sleeping bag at the rudimentary field station and heated a can of beans on an open fire for his dinner.

From those painstaking efforts would rise new varieties of wheat, strains resistant to deadly fungal diseases, that would thrive in the varied Mexican climates, in poor, overworked soils in the teeming lands of India and Pakistan, and in family plots throughout much of the world. From the work of this man, a farm boy named Norman Borlaug, would come a green revolution, one that would feed the world for at least a little while and change forever how we think about the division between nature and man.

The Early Years: Kidnapped

Norman Borlaug was born in 1914 in the hamlet of Saude, Iowa, on his grandparents' farm. The great-grandson of Norwegian immigrants and the oldest of four children, Borlaug learned early on the meaning of hard work. From age seven, he was an active participant on the farm, tilling the fields, milking the cows, feeding the livestock, mucking out barns and splitting wood. He trekked a mile-and-a-half in the winter, in blowing snow, to and from a one-room county schoolhouse, often arriving home with toes, fingers and nose numb.

But Borlaug never thought much of it—he knew that all boys his age in his county worked just as hard. Unlike most boys, however, Borlaug's family had greater ambitions for him than simply plowing fields. As soon as the boy could understand, his grandfather Nels drilled one over-riding ambition into his grandson: "Get an education." His grandfather had received only three years of formal schooling, but retained a deep longing to learn. "You're wiser to fill your head now if you want to fill your belly later on," he told his oldest grandchild. And though he wasn't the greatest shakes as a student, Borlaug had, as his second cousin and eighth-grade teacher Sina Borlaug said, "grit." She recommended he go on to the Cresco High School, fifteen arduous miles from the Borlaug homestead. The added learning, she said, would "make him."

She was right. Cresco High School became instrumental in Borlaug's life, and not just for the education it provided. In high school, Borlaug was introduced to the world of athletics, eventually settling on wrestling as his sport. It was during his wrestling practices that Borlaug, coached by his high school teacher, David Bartelma, who had gone to the 1924 Olympic games, learned many of the vital lessons that would remain with him throughout his life. For instance, don't just try to overpower your opponent—outthink him at the same time as you act. And, most importantly, "Give the best that God gave you. If you won't do that, don't bother to compete."

The golden days of Borlaug's youth were abruptly shattered, however, with the stock market crash of 1929 and the Great Depression that followed; yet despite the tremendous financial struggle, the Borlaug family scrimped together enough to allow Norman to finish high school. He graduated in 1932, at a time when the country had no jobs to offer and his family no money to pay for college.

But Borlaug was nothing if not determined. He learned of a scholarship to train a science teacher at the Iowa State Teacher's College in Cedar Falls, Iowa. There was only one problem: It wouldn't become available for another year. Borlaug spent that year cutting fence posts and hunting and trapping animals for their meat and pelts. He hired himself out to neighboring farms for spring planting and fall harvesting for the princely sum of fifty cents a day, and considered himself rich for the salary. Every penny counted; every penny was one more step on his road to college. By summer he had squirreled away $50, and soon thereafter he learned he had been accepted to the college with a partial grant.

Then, as would happen so often throughout his life, fate intervened. A week before he was due to leave for Cedar Falls, Borlaug was kidnapped by a young man who had graduated from Cresco High School two years before and now attended the University of Minnesota. George Champlin had heard about Borlaug from Norman's old wrestling coach, Dave Bartelma. He turned up at the Borlaug door to whisk the former wrestler off to Minneapolis and the university's fledgling wrestling team. Here was Borlaug's chance to attend a full university rather than a two-year teacher's college. This was his opportunity to receive the rich higher education he and his grandfather had always dreamed of.

How could he say no?

With $61 in his pocket, thanks to $11 his grandfather added to his existing stash, and the mere promise of a job, Borlaug hopped into Champlin's car, made a stop to pick up his high school friend and fellow wrestler Erwin Upton who, by way of Champlin's recruiting, had also been lured to Minnesota, and headed 150 miles away to the big city.

The True Face of Hunger

The high spirits Borlaug enjoyed during the drive to Minneapolis were dashed during his first week in Minneapolis. Because of differences in the high school education offered in Iowa and Minnesota, he and Upton were required to take an entrance exam. Upton passed easily; Borlaug failed. There was one option left: The university offered him a place at its "General College," which some snide students called the "college for dumbbells." Borlaug could take a broad sweep of subjects and at the end of two years receive a degree. It wasn't what he'd hoped and dreamed of, but he had no other choice. Despite his anger and shame at having flunked the exam, he agreed to enter.

Even before classes started, Champlin made good on his promise to help the two young men find jobs. They worked in a local coffee shop, serving breakfast, lunch, and dinner. Their only pay was sitting down to eat those meals. It was during those meals that Borlaug met a pert, pretty young woman named Margaret Gibson, a 20-year-old sophomore in the College of Education. The two quickly bonded, brought together by their similar rural upbringings and their ability to talk about anything. One topic they spoke of often during that first semester was Borlaug's frustration with the remedial college he attended. "I'm nothing special," he told Margaret one night, "but I know damn well I'm better than that. That place for the misfits they expect to drop out."

Then get out, she told him. Work harder.

So, as he'd done on the wrestling mat, in the fields, while hunting, and in whatever he did, Borlaug set a goal, focused on it, and allowed absolutely nothing to distract him. He worked, as his early biographer Leonard Bickel wrote in a 1974 book, "with an intensity that attracted attention." By the end of the first semester, he convinced the administrators to let him out of the general college and into the College of Agriculture to major in forestry.

Two incidents marked Borlaug during his first year at Minnesota. One occurred soon after he arrived in the metropolis of Minneapolis, when he went to visit the university's agricultural campus on the outskirts of St. Paul. Walking back into town, he happened upon a factory riot: The bad feelings were instigated when managers slashed workers' wages in half, and then were fueled when the workers struck. The desperation he saw in the workers' faces as they fought against truncheons with their bare hands, fighting to feed their families, stayed with him throughout his life and ignited in him a burning desire to eradicate hunger.

The second incident affected him more directly. As a member of the wrestling team, Borlaug often had to fast to lose weight so he could compete in his weight class. One time he went four days without food and with very little water, spending hours in a steam box to sweat away the pounds. On his fifth day without a meal, frustrated when he discovered he was still a pound overweight, the normally gentle Borlaug lashed out at another wrestler and would have punched him in the face if his friends hadn't pulled him away. For the usually calm Borlaug, such an outburst was, to say the least, uncharacteristic. As he told Margaret that night: "I think I've learned a primal rule of nature. You see, it wasn't me at all. It was primitive, rudimentary. I can't explain how hungry I was. I was starving, and I found out that a hungry man is worse than a hungry beast."

In December 1935, Margaret dropped out of school to get a job. The scrimping and saving were just too much, she told Borlaug: "I cannot tolerate it any longer." She got a job as a proofreader and settled into living in a one-room apartment. By now, the two had an unannounced agreement to eventually get married, when they could afford it. Pennies still counted during those Depression days, and Borlaug spent summers working for the U.S. Forest Service watching for fires, alone on Cold Mountain in the Idaho National Forest, the most remote wilderness in the lower forty-eight states. During the school year, he was helped by the National Youth Administration program, earning fifteen cents an hour for part-time jobs as credit towards tuition.

Then, as he prepared to start his final year of college in 1937, the economic tide seemed to turn. Borlaug was offered a permanent job with the Forest Service as an assistant ranger in the Idaho National Forest once he graduated. Elated, he proposed to Margaret, who immediately said "Yes." A few days later, on September 27, the couple were married in Margaret's brother's sitting room, with a few family members in attendance. There was no lavish reception, no honeymoon, just a return to her one-room apartment with the bath down the hall.

But a few days before Christmas and his graduation, Borlaug received a letter that changed everything. Due to budget cuts, his job no longer existed. It would turn out to be the best thing that could have possibly happened to him and millions of people around the world.

From Wrestler to Protégé

During his final undergraduate semester, Borlaug had had a strange encounter with the head of the plant pathology department, E.C. Stakman, one of the most respected scientists in the field. For years, Borlaug thought the interaction was accidental; yet Stakman, who had seen the courage and tenacity Borlaug showed during a wrestling match, had deliberately sought out the young man to test his intelligence. Stakman walked into the forest pathology lab where Borlaug was examining wood samples under a microscope and, without introducing himself, began peppering the student with questions about the samples. This Ph.D.-level pop quiz, Borlaug later learned, was designed to test his ability to meet challenges head-on. Borlaug passed with flying colors.

A few weeks later, Borlaug attended a lecture Stakman delivered on his decades-long research into the challenges of rust diseases in wheat. The parasitic fungi causing the disease—the bane of all farmers—were capable of changing their genetic makeup in less than a year to attack previously resistant crops, wiping out a season's yield in days. "Rust diseases are the relentless, voracious destroyers of man's food," Stakman told his audience, "and we must fight them by all means open to science." Stakman didn't stop there, however. He warned of another threat to the world's food supply—man himself. Population growth would soon outstrip food supply, he said, and the one thing that could reverse the trend was science. Even there, however, his message held foreboding: "Do not deceive yourselves—ever—that the scientific approach is omnipotent. It will make its mistakes, but it will take us further than has ever been possible to eradicate the miseries of hunger and starvation from this earth." If ever a speech was designed to ignite and motivate, this was it. "One day I would like to go back and study under that man, if it is ever possible," Borlaug told Margaret that night.

When the implications of the forestry letter sank in, Margaret gave him his chance. Go to graduate school with Stakman, she urged her husband. They would somehow manage on the income from her proof-reading, which she loved doing.

Buoyed by her support, Borlaug walked into Stakman's office the next morning and told the man he wanted to spend a couple of months doing graduate work in forest pathology. He might just as well have told Stakman he was ready to begin painting his face and touring with the circus. No way, Stakman told the young man. You don't dip your toe into the world of graduate work—you jump in. And forest pathology is self-limiting: Focus on all plant pathology and you can work with any species.

Borlaug became a Stakman disciple. He received his master's degree in 1940 and went on to study for his Ph.D., also under Stakman. In late 1941, as Borlaug was completing his thesis, Stakman arranged for his protégé's first professional job. Upon graduation, Borlaug was to work for the chemical company DuPont in Wilmington, Delaware, in its biochemical laboratory group. His salary would be $2,800 a year. Borlaug accepted, and in early 1940 he and Margaret bought their first car, a secondhand Pontiac, and moved to Wilmington.

To Mexico

Borlaug had barely begun his professional life when the United States entered World War II. Although he wanted to enlist after the attack on Pearl Harbor, he was told his work at DuPont was considered "too valuable" to the war effort. So Borlaug spent the war years at DuPont, where he and his colleagues developed, among other products, camouflage paint, aerosols, and chemicals to purify water. He also oversaw the mass production of a new insecticide: DDT. During this time he became a father. His daughter, Norma Jean (Jeannie), was born on September 27, 1943.

Then the Rockefeller Foundation came calling. In 1940, the Rockefeller Foundation sent a delegation of agricultural experts, including Stakman, to tour Mexico and report on what was required to help the starving country develop a viable agricultural program. Their recommendation was to create an agricultural research infrastructure. It was the only way, the men reported, that Mexico would ever be able to feed its own people without depending on the handouts of richer countries like the United States. The foundation agreed to fund the project and asked Stakman to pick the man to lead it. Stakman recommended one of his former students, George Harrar. Together, he and Harrar hammered out a plan for Mexico: they would create an Office of Special Studies that Harrar would run in cooperation with the Mexican government. But the Americans were not there to take over. A major goal of the office was to train young Mexican scientists in modern farming techniques so they could run the venture, sooner rather than later. It was a new twist on the "teach a man to fish" parable.

When Harrar needed a plant pathologist for his team, Stakman recommended his protégé, Norman Borlaug. "He has great depth of courage and determination," he wrote to the Rockefeller board. "He will not be defeated by difficulty and he burns with a missionary zeal."

Borlaug was intrigued. The idea of improving the quality of life of a whole country was attractive. As Richard Zeyen, professor of plant pathology at the University of Minnesota, who has known Borlaug for decades, explains: "He had that small town farmers-help-farmers-help-each-other spirit. He was strong and he knew he was strong and he knew he could use his strength to help others." In 1944, thanks to intervention from the Rockefeller people, his wartime designation was lifted and he was free to go to Mexico. Even though Margaret was pregnant again and Mexico was 2,000 miles away and still primitive, Borlaug agreed to go.

He drove from Wilmington to Laredo, Texas, where he met up with Edwin Wellhausen, the communications specialist on the project. From there they made the three-day, 800-mile journey through the dusty, hot, yet colorful landscape to the agricultural school at Chapingo, twenty miles outside Mexico City. There, Borlaug found a newly built adobe shed with a tarpaper roof surrounded by 800 acres of weed-choked fields. There were no trucks or tractors, no research materials, no irrigation supplies, no gasoline or tires. The war had taken everything; nothing was left for Mexico. In addition, the Americans had to deal with the frustrating Mexican bureaucracy, which Wellhausen described as similar to "punching a featherbed." As he went to sleep that first night in his Mexico City hotel, Borlaug wondered what the future held for him in this strange country.

The three principals at Chapingo—Borlaug, Wellhausen, and a soil expert—set to work. They had to do everything from developing roads and underground irrigation to readying the land for planting. To speed things along, they tried to get farmers in the surrounding towns to plant small experimental patches; but the scientists quickly realized that the farmers were too poor, hungry, and illiterate to take on such a role. Things only got more depressing from there.

Margaret called in early November to say that their baby had been born with severe spina bifida, a condition in which the spinal column does not close properly. Borlaug flew back to Wilmington to see his son, Scotty, for the first and last time. The infant could not leave the hospital and was too fragile to hold. Borlaug asked Margaret if she wanted him to look for work locally, but she said no; doctors had told her there was no hope for Scotty and she knew her husband's heart was in Mexico. Margaret stayed behind with the baby, but when it was clear there was no hope, she followed her doctor's advice and, reluctantly, took Jeanie and rode the train to Mexico City, leaving Scotty behind. He died a few months later. It would be four more years before the Borlaugs had another son, born in Mexico City, whom they would name Bill.

Radically Speeding Up Agricultural Time

The science of agriculture, unlike that of some other disciplines, has few "ah-ha!" moments. There are almost no major moments of discovery, no miraculous recoveries, no sudden breakthroughs. There are only the slow, inexorable seasons, the centuries-old methods of preparing the soil, planting the seed, and observing the traits of what grows.

When you're trying to feed a nation, however, such a pace can be tortuous, and so, out of his need for action and his own frustration, Norman Borlaug sped ahead; in doing so, he forced the rest of agricultural science to keep up. To find the right types of wheat that would flourish throughout Mexico, he knew he would have to cross thousands of varieties, searching for the plants whose mutations would provide increased yield while also fighting off the fungi causing rust. If he did it the old way, it could take decades of crossing the wheat, planting the new varieties, and waiting to see what happened. But the desperate need he saw in Mexico impelled him to find a way to speed up agricultural time.

The first of three great innovations Borlaug developed in Mexico was shuttle breeding. Rather than planting a wheat crop, waiting for it to grow, and harvesting it to see which varieties survived, he came up with the idea of supercharging the process by growing two plantings in the same year. The summer crop would grow in the poor, high-altitude, bone-dry soil of the Chapingo region outside Mexico City, and a winter crop would grow over 1,200 miles north in the irrigated sea-level Yaqui Valley in Sonora, with its more fertile soil and better growing conditions. The differences in altitude and temperature meant the two areas had different growing seasons, so Borlaug could not only grow two crops of wheat in a year, he could also see if the plants that grew well in one region also worked in the other. He would harvest seeds in the summer from Chapingo and plant them in the winter in Sonora, and then harvest seeds from the winter's Sonora crop for planting in the Chapingo summer.

This was radically different from how agricultural science was done at the time. His colleagues thought he was nuts. Harrar didn't want to expend resources in two areas of the country. "It makes no sense to risk a round trip of 2,500 miles through that country," Harrar said, knowing how hard it was to travel overland in Mexico at the time. "You could lose everything—and for what? The guts of our problem is here, in the poverty areas. You've got to get that clear!" Harrar enlisted Ed Wellhausen to dissuade Borlaug.

But, as Bickel related in his biography, Borlaug thrust his chin out and knitted his brows together. His stance was reminiscent of the powerful wrestler he once was, only this time his opponent was not another young man, but his boss, and at stake wasn't a wrestling match, but the stomachs and lives of millions of Mexicans.

"Don't try to discourage me, Ed. I know how much work is involved. Don't tell me what can't be done. Tell me what needs to be done—and let me do it. There's one single factor that makes the Yaqui effort worth a try and that's rust. Breeding two generations a year means beating and staying ahead of the shifty stem-rust organism. If I can lick that problem by working in Sonora, then we've won a victory. To hell with the extra work and strain. It's got to be done, and I believe I can do it."

Harrar relented and Borlaug managed to get one season's wheat planted. But a visit the following year by the three-man Rockefeller Foundation team, including Stakman, brought the whole issue to the fore again.

"You're going to be going in circles with such a program," one of the visitors, plant geneticist Dr. H. K. Hayes, warned him, "One step forward, the next step backward." Hayes was of the old-school belief that plants must be bred where they will be grown; it was thought that there were too many local variables that could affect the plants to do otherwise.

The foundation team voted against the Sonora planting and Harrar testily told Borlaug, "I have told you each time the subject comes up that you should concentrate your efforts in the Bajio area [where Borlaug was based]. This scheme of yours has been considered twice, at your insistence, and voted down twice. Why can't you accept that?"

"If this is a firm decision, I also make a firm decision. You will have to find someone else to conform to your rules," Borlaug told him. "You're laying down a policy that is wrong. And I can't go along with it. As of now, I resign. You'll have it in writing first thing in the morning."

And with that, he stalked out of the room.

The next morning Stakman convinced Borlaug to go on to work at the Chapingo station, promising him he could turn in his resignation letter at the end of the day. Stakman knew that Borlaug's plan might just work, so he went to see Harrar. Coincidentally, that very day Harrar received a letter from a farmer in Sonora praising Borlaug and the Rockefeller Foundation for planting new strains of disease-resistant wheat on his land. "I want to say what is happening here with Dr. Borlaug will have a tremendous effect within a short time," the farmer's letter stated. After a discussion, no one knows how much influenced by the farmer's letter, Harrar agreed to allow Borlaug to continue shuttling his wheat and himself between the two regions.

The second major innovation Borlaug brought to Mexico was high-volume crossbreeding. At the time, plant breeders typically only crossed a few plants each season, waiting until the plants were harvested before choosing the best varieties to use for crossbreeding a few more varieties the following year. With this method, it could take decades before a viable new breed emerged.

Ever impatient, Borlaug took a different approach. Knowing he had one chance in a thousand to hit upon a winning variety, he collected thousands of wheat varieties from throughout the world and began crossing them simultaneously, hoping to find those that were most resistant to fungal diseases like rust and could best survive in the various Mexican soils and climates. He also optimized multiline breeding, backcrossing hybrids with a single parent, thus putting multiple disease fighting genes into a variety. Hence, the long hours spent under the broiling Mexican sun hand-pollinating his wheat varieties. And when he finished on Friday evenings he would get in his car and drive the six hours to Mexico City to spend the weekend with his family. Saturday mornings he got up to coach his son's baseball team. It played in Mexico's very first Little League, one that he and Dr. John Niederhauser, another Rockefeller scientist, started. As a colleague later wrote of him during this time: "Work was not just a word to him; it became a code of honor. If genius is 'an infinite capacity for taking pains,' Borlaug had it."

Most importantly, he never sought perfection, only something better. "We don't have time to wait for perfection," he said, thinking of the starving population. To Borlaug, 40 percent better and harvestable this year was better than 90 percent better in five years. By 1952, he had more than 40,000 wheat varieties in his nurseries and more than 6,000 individual crosses—all meticulously recorded. Combined with the shuttle-breeding approach, his approach cut the time required to develop new varieties in half. By 1956, Borlaug had developed forty new rust-resistant strains of tall wheat. There was only one problem, and it was a major one: When the farmers began using the large amounts of fertilizer required to increase yields, the wheat grew even taller and began lodging, which is a farmer's way of saying that the wheat had a tendency to flop over in wind and rain, potentially ruining the crop.

So, Borlaug embarked on his third major innovation. He crossed his rust-resistant varieties with a new Japanese dwarf variety of wheat to create a shorter, stiffer variety of wheat. The results were spectacular—the dwarf Mexican wheat varieties doubled the country's yield from about two tons per acre to just over four tons per acre.

On to India and the Start of the Green Revolution

By the late 1950s, Borlaug's wheat was so successful that Mexico was self-sufficient in wheat production and had its own scientists to run agricultural projects. It was time for Borlaug to move on and he began negotiating a job with a tropical fruit company to investigate banana diseases. Meanwhile, the Rockefeller Foundation found a new crisis to focus upon: impending famine in Asia.

In the early 1960s, Asia was undergoing a population explosion and its farmers were not producing enough food to keep up. Experts began predicting mass starvation. Some were actually calling for the United States to pull out of India and let millions of people starve in a social engineering attempt to reduce its population. Bestselling books came out, including the Paddock brothers' *Famine 1975! America's Decision: Who will Survive?*, which predicted that the United State's food export programs could not keep up and policy makers would have to choose who would live and who would die. *The Population Bomb* made its author, Paul Ehrlich, famous. He proclaimed that it was a "fantasy" that India would ever feed itself, and that "in the 1970s and 1980s, hundreds of millions of people will starve to death in spite of any crash programs embarked upon now."

It was in the midst of this crisis that the foundation deployed Borlaug and other scientists on a fact-finding mission to the region. The misery, poverty, and hunger the delegation saw there left many of them shaken; but Norman Borlaug was not a defeatist. He attacked problems. In this case, he came up with a scheme he called the Kick-Off Approach that would work on three sets of factors—technical, psychological, and economic—to overhaul the agricultural systems. The technical work began immediately. In 1963, the Rockefeller Foundation created the International Maize and Wheat Improvement Center in Mexico, known by its Spanish acronym CIMMYT. Trainees arrived in Mexico from Afghanistan, Cyprus, Saudi Arabia, Libya, Pakistan, and various other Middle Eastern countries, as well as from ten South American countries, to learn every aspect of agricultural development. Conspicuously absent was anyone from India, whose agricultural policy makers were suspicious of the whole endeavor. Simultaneously, Borlaug established nurseries in North Africa, the Middle East, South Asia, and throughout Latin America to find the best possible wheat varieties from the Mexican strains for use in these areas. Once the trainees learned all they could in Mexico, they returned to their native countries to manage the nurseries and oversee the dissemination of the best varieties.

Borlaug's second tactic evolved from his understanding of farmers' psychology. He knew that the massive famines in the Soviet Union in 1932 and China in 1958 had been partly the result of top-down government policy. While he believed government policy was integral, he understood that agricultural advances always begin within the minds of individual farmers. "He started with the farmers, always the farmers," explains Richard Zeyen. In Pakistan, for instance, Borlaug planted native wheat in plots next to the new wheat. "Any fool could see the difference," says Zeyen. "Then he'd invite the farmers in and they would get very excited. But Borlaug would say about the new varieties, 'this is like gold. Don't touch it!' Of course, as soon as his back was turned, they were taking the heads off the wheat and gathering the seed." In this way, Borlaug built demand for his new wheat from the ground up, encouraging farmers to put grassroots pressure on bureaucrats to find ways to support them. "Farmers are very intelligent people," explains Zeyen. "They knew that if they could get more yield on the same land with these new varieties of plants, then they could have more money and a better family life."

Borlaug's third approach required governments to radically change their agricultural policies. Because his wheat required new methods of farming, farmers needed access to a system of credit so they could afford fertilizer and seed in advance of planting. The fertilizer was always a fight. Borlaug's wheat was bred to take up heavy applications of fertilizer and economists at the time did not realize that fertilizer was an investment that would be repaid with much higher yields. Also, Borlaug insisted that government money be available for drainage and irrigation improvements, as well as equipment to plant and harvest the wheat. In addition, farmers needed the government to create an infrastructure of roads, bridges, and warehouses to get the wheat from the farms to market. Finally, governments had to guarantee farmers fair prices for their future crops in advance. This American showing up on their doorstep and telling them how to run their agricultural policies was often met with great suspicion by governmental officials, but

Borlaug knew if he merely provided them with academic papers nothing would change. He had to find a way to impress upon officials that the miracle seed was only the first step to a fruitful agricultural policy.

This was especially difficult in India, which had become independent from Great Britain in 1947. In 1964, the country's first prime minister had died and the fragile democracy was teetering between the Cold War superpowers. There was a very real possibility it would tip towards the Soviet Union. To make matters worse, India's population was starving, thanks to a failed focus on heavy industry instead of agriculture. To show its support for India (and to unload its grain surplus), the U.S. had started the massive Food for Peace program in the early 1960s, shipping 5 million tons of emergency wheat grain in 1964 and 1965. Canada and Australia also sent wheat in what became the largest food rescue operation in history. But it was like trying to put out a forest fire with a bucket brigade.

As Zeyen recalls, "We were going broke trying to feed these people." It was critical that India learn to feed itself. That meant moving the technological advancements identified in Mexico to the Asian country. There was just one problem—India didn't want it. Not only were officials suspicious of non-native wheat, they also worried that feeding the masses would upset the fragile social order and government by the elite.

Since India never sent trainees to Mexico in those early years, Borlaug and the Rockefeller Foundation began showing India how agriculture reform was progressing in India's bitter rival Pakistan. It was in Pakistan that Leon Hesser, the author of Borlaug's 2006 biography, *The Man Who Fed the World*, first met his subject forty years earlier. Hesser was working for the U.S. State Department and was charged with increasing food production in Pakistan, an impossible task until Borlaug showed up. "He simply came to my rescue because at that time my team and I didn't have a lot to extend to the Pakistani farmers," Hesser recalls. "What Borlaug did was a real miracle." As Zeyen appraised the situation, "Once a country like Pakistan took off and started heading into self-sufficiency, India had no choice but to participate."

But when India finally expressed interest, it wanted all the results without the necessary policy changes. Once again, Borlaug was forced to use his wrestler's tenacity. Borlaug had already impressed the Minister of Agriculture, Dr. Shri C. Subramaniam, with his program. Unfortunately, Subramaniam had been voted out of government. As a man with deep concern for his countrymen, he arranged for Borlaug to meet the man in India's government who had the power to change agricultural policy, Deputy Prime Minister Ashoka Mehta, the number two man in India's government. Subramaniam told Borlaug to be frank and explicit about what would be required to change Indian farming. Borlaug prepared himself, but he knew it was a long shot attempt; he confided in a colleague, "I am going to follow Minister Subramaniam's suggestion and speak very bluntly to Minister Mehta about the government's disastrous policy on fertilizer, credit, and grain prices. The meeting will very likely be stormy and I may be asked to leave the country, so you better keep a low profile. Should that happen, you and Dr. Anderson can keep the wheat revolution moving forward."

The meeting began quietly but escalated into a shouting match when Borlaug insisted that the government provide fertilizer. "For several minutes there was chaos, with both of us talking in loud voices at the same time. A flood of loud, angry words was emitted by both of us until we both ran out of breath and began to talk in rational tones once again," Borlaug recalled.

Borlaug attacked every argument the minister made for slow change. He insisted that India import more fertilizer and build fertilizer plants, and he provided a list of international agencies that could lend the money to finance these operations, but none of this was acceptable to Mr. Mehta. But then Borlaug told Mehta of the farmers: If the new policies were adopted, the farmers would dramatically increase production and this would stimulate the whole economy. Borlaug believed India could feed its people. "As the meeting ended, I think we had re-established mutual respect, if not mutual friendship," Borlaug related.

Two weeks later, India's newspapers were full of stories concerning India's new agricultural policy. The government had capitulated—Borlaug had won—and the farmers could go to work. That year Pakistan and India placed an order for 600 tons of Mexican wheat seed. It took a convoy of thirty-five trucks to transport the seed from Mexico City to the port of Los Angeles, where they were held up by the raging Watts riots. At the same time, the bank called: Pakistan's $95,000 draft to pay for the seed had several misspellings and the bank wouldn't accept it. Borlaug, never a man comfortable with bureaucracy, finally lost his temper. "Get that wheat aboard the ship and send it on its damn way!" he yelled. The wheat was loaded and the ship sailed, but a few days later war broke out between India and Pakistan.

The seed made it to the subcontinent and farmers, sometimes in the shadow of artillery and within the earshot of gunfire, sowed it in their fields. But it was planted late and much of it had been damaged back in Mexico when it was fumigated, so little of it germinated. Borlaug flew over to view the disaster. He had a thought: "Feed them," he commanded, "feed the seeds." The farmers tripled the fertilizer application and the crop responded. Despite its slow start, the first crop still produced yields 70 percent higher than India had been producing on its own, and the next crop showed a 98 percent improvement.

Despite this relatively small start and even with the mountains of grain being shipped by Western countries, not enough was being done. "We'll not cure the damn trouble by filling food bowls," Borlaug warned the Indian agricultural minister. "If this is not to happen again, you will have to fill your fields with high-yield dwarf wheat along with new dwarf rice varieties being produced in the Philippines." India's government finally moved boldly. It ordered 18,000 tons of Mexican wheat seed, chartered two freighters, and had the largest seed purchase in history shipped directly from Mexico across the Pacific to India. The following year, Pakistan ordered 42,000 tons of seed. It was as if they were in a race to succeed. By 1968, Pakistan was self-sufficient in grain production; by 1974, India joined it. In the years that followed, caloric intake per person steadily rose as wheat production rose, and the percentage of undernourished people in India declined from 39 percent to 22 percent. China, watching its rival and neighbor, also changed its agricultural policies and saw the proportion of undernourished people in their country drop from 52 percent to 12 percent over the next three decades. Perhaps just as impressive was that the increased production came with no increase in the amount of land required to grow the crops. Yields increased in India by 150 percent and in China by 300 percent. India practiced double-cropping, a system in which two crops a season were planted, one watered by the natural monsoon, the other by the artificial monsoon of irrigation (now possible thanks to the enormous irrigation facilities and dams the government had constructed). Without it, experts predicted, the country would have had to plow under another 100 million acres of land, an area about the size of California.

In 1968, United States Agency for International Development director William Gaud stated, "The rapid spread of modern wheat and rice varieties throughout Asia and other developments in the field of agriculture contain the makings of a new revolution—I call it a Green Revolution—based on the application of science and technology." It was the first time the phrase was used, but it stuck because the effects were so monumental. Thomas R. DeGregori, Professor of Economics at Houston University, says, "At the core of the Green Revolution was a grain revolution, with Borlaug's wheat providing roughly 23 percent of the world's calories." Borlaug was not alone in making these dramatic agricultural advances, but it was his extensive plant breeding in Mexico and his iron-willed drive to make government officials take action that was the catalyst for much of the Green Revolution.

It Is Nutrition that Saves Lives

Although in the 1960s, the so-called experts were predicting mass starvation by the middle of the 1970s, more realistic assessments of food supply and population projections from the time show that such dire circumstances were unlikely to occur even without the Green Revolution's increase in productivity. Scholars such as Amartya Sen, a Nobel Prize winner in economics, have shown that famine due to lack of food was mostly eliminated in the twentieth century. Modern civilization had both the will and the capability to transport enormous quantities of food anywhere in the world to avert mass starvation. The real cause of modern famines was distribution failure caused by war, unreliable government policies, or economic problems. While Borlaug's revolution may not have saved those such as the tiny children with bloated stomachs whose image most often comes to mind when famines occur, he did save million upon millions of lives.

The long-term death toll from food shortages stems much more from undernourishment than outright starvation. From the late 1800s, it was known that nutrition is important in warding off specific diseases such as scurvy. But two decades ago, Nobel laureate Robert William Fogel, among others, discovered that nutrition also had a dramatic effect on the length of a person's life. In other words, your mother telling you to eat your veggies as a child actually has an effect on how long you live. His studies showed that 20 percent of the population in pre-industrial Europe didn't consume enough calories to even fuel a full day's work. However, caloric fuel was only the most obvious benefit of food—Fogel also discussed studies by researchers who found an interesting correlation between the heights of Union Army veterans of the Civil War and their deaths. Records were kept of these men when they joined the Army and throughout their lives as they received veterans' benefits. It was found that the taller the soldier was, the longer he lived. When researchers examined other populations, similar findings corroborated this relationship. Although height is partly determined by genetics, there is a component that is influenced by environment, especially childhood nutrition. In fact, it has been found that average height has increased dramatically in populations around the globe whenever nutrition has improved. For example, the average Frenchman in 1775 was five feet, four inches tall and is now five feet, ten inches tall. Additional height is an indicator of more robust organs that better stave off disease.

Other researchers have found equally interesting correlations. Gabriele Doblhammer and James W. Vaupel discovered that people who are born in the months of October, November, and December live a half-year longer on average than those born in other months. When this study was repeated in the Southern Hemisphere, the same was found to be true, but for months that were six months out of phase. By ruling out other possible causes, the researchers determined that the difference was linked to the seasonal availability of fruits and vegetables to mothers when they were pregnant. So it isn't only what you eat, it's also what your mother eats while pregnant that influences how long you will live.

Recently, new experiments in epigenetics, a revolutionary field involving DNA inheritance, have shown that poor nutrition can impact one's chromosomes such that effects can even be passed on to future generations. Epigenetics involves the study not of the makeup of the DNA molecule itself, but of the chemicals and structures that affect how the genes coded by DNA are expressed. These structures can be affected in a long-term way by environmental influences such as diet. The offspring of some mice can be made healthier and longer-lived simply by changing their parents' diet. In nematodes, the effects of an individual's environment can remain in their descendant's genetic code for many generations. If the same is true in humans, what your grandparents ate might be influencing your current health.

These are examples of how nutrition affects an adult's lifespan; the effects of malnutrition on children are more severe. Estimates today are that half of all children who die in the developing world do so because of malnutrition. Measuring growth retardation, or stunting, is one way scientists assess the health of children. Stunting is a direct consequence of poor nutrition and is strongly correlated with increased mortality. In 1980, just after the start of the Green Revolution, 47 percent of all children in the developing world were stunted. By the year 2000, this figure had dropped to 33 percent. A child who is only mildly malnourished has twice the chance of dying from childhood diseases as a child who is well-nourished; one with severe malnutrition has an eight-fold chance of dying. Children's small, developing bodies are simply not robust enough to fight off common childhood infections.

The Green Revolution laid the cornerstone for adequate nourishment by increasing the available calories and protein of the developing world's people. Since what people eat, what their mothers ate, and possibly even what their grandparents ate all affect how long people live (and whether children even live to grow up), the impact of Norman Borlaug's Green Revolution on saving people's lives has been profound and is still being played out. Our estimates are that Norman Borlaug's Green Revolution saved over 245 million lives due to improved nutrition. The most significant measurable change was the decline in the death rate of children under the age of five. In addition, the increased nutrition has allowed countless millions of adults to live gracefully into old age.

The Nobel Prize

It was October 20, 1970, and Norman Borlaug was doing what he had done at this time of year for the past twenty-six years—standing in a Mexican wheat plot dressed in mud-splattered clothes, boots, and a baseball cap, choosing exceptional wheat varieties. Around 10 AM, he heard the sound of a car bumping along the rutted road at the edge of the Toluca station. When it stopped and he saw his wife emerge, he became frightened, certain something must have happened to one of their two children. He could never have imagined the news she'd come to tell him.

"What's wrong?" he cried, dropping his wheat samples and running towards her.

"Nothing," she laughed. "You've won the Nobel Peace Prize, that's all."

At first, he refused to believe it. Borlaug insisted Margaret return to their house; he still had a day's work to accomplish. As it turned out, he had about forty more minutes of toil before the press descended upon him.

What, you might ask, do wheat and bread have to do with peace? Perhaps expecting that question, the Nobel Committee answered it when it awarded Borlaug the 1970 Peace prize. The committee compared Borlaug's work "to the basic human right of freedom from starvation as recognized by the Charter of the United Nations," and declared that his work had "helped to turn pessimism into optimism in the dramatic race between population explosion and food production."

For Borlaug, however, the prize was a complete surprise. "You have to understand that Norman Borlaug has no ego," Zeyen says. "He's the world's greatest humanist. He cannot stand to see people suffer." Borlaug took his entire family with him for the ceremonies in Oslo, Norway (where the Peace prize is awarded; the other Nobel Prizes are awarded in Sweden), and gave a powerful Nobel lecture in which he warned that the continuing race to feed the world's population was anything but over. "Almost certainly," he said, "the first essential component of social justice is adequate food for all mankind. Food is the moral right of all who are born into this world. Yet today 50 percent of the world's population goes hungry. Without food, man can live at most but a few weeks; without it, all other components of social justice are meaningless .. . . If you desire peace, cultivate justice, but at the same time cultivate the fields to produce more bread; otherwise there will be no peace."

He also insisted the Green Revolution was not complete. "Perhaps the term Green Revolution as commonly used is premature, too optimistic, or too broad in scope," he said. "The reality is that only some crops have been modified; only some farmers have benefited. And most of the benefits have come in irrigated areas."

An Environmentalist

As Borlaug continued his efforts to expand agricultural success, he found himself fighting off critics who began denouncing his methods, particularly the large quantities of fertilizers and pesticides required to achieve the high yields. Damaging pesticides can run off the fields into streams and rivers; this also occurs with surplus fertilizer, creating algal blooms that remove oxygen from water. An example is the Dead Zone, an area mostly lacking in life that begins at the mouth of the Mississippi River and extends far out into the Gulf of Mexico.

As Borlaug told his biographer Leon Hesser: "I was at once drawn into the vortex. As soon as you start to challenge an emotive issue you get attacked by feverish, committed people. I was subjected to insult, mudslinging, trashy verbal assault. I knew all that would come. But what could I do but accept the responsibility?"

As a man of action in the midst of crises, Borlaug has said, "Some of the environmental lobbyists of the Western nations are the salt of the earth, but many of them are elitists. They've never experienced the physical sensation of hunger. They do their lobbying from comfortable office suites in Washington or Brussels. They have never produced a ton of food. If they lived just one month amid the misery of the developing world, as I have for sixty years, they'd be crying out for fertilizer, herbicides, irrigation canals, and tractors and be outraged that fashionable elitists back home were trying to deny them these things."

Recall that Borlaug worked alone in the wilderness in Idaho during his college summers. "To this day," Borlaug has said, "I enjoy nature, the luxury of undisturbed wilderness, forests, mountains, lakes, rivers, and deserts and their wildlife. But I also know that the greatest danger to their perpetuity is the pressure of human population." In a paper he wrote in 2000, he said, "We all owe a debt of gratitude to the environmental movement that has taken place over the past forty years. This movement has led to legislation to improve air and water quality, protect wildlife, control the disposal of toxic wastes, protect the soils, and reduce the loss of biodiversity. It is ironic, therefore, that the platform of the antibiotechnology extremists, if it were to be adopted, would have grievous consequences for both the environment and humanity. I often ask the critics of modern agricultural technology: What would the world have been like without the technological advances that have occurred? For those who profess a concern for protecting the environment, consider the positive impact resulting from the application of science-based technology."

Borlaug points out that in order to produce 1999's world cereal crop using 1961 agricultural methods, an additional 2 billion acres would have to be under cultivation—that's 3 million square miles, about the size of the contiguous United States. Instead, this land can be used for other purposes, such as wilderness preserves. As Vaclav Smil of the University of Manitoba says, "Without the 80 million tons of nitrogen consumed annually (from chemical fertilizer), the world could sustain no more than four billion people, two billion fewer than inhabit the Earth today."

This type of evidence suggests that Borlaug's Green Revolution may have saved more land for wilderness than any single environmental organization.

Workaholism Redefined

The Nobel Peace Prize slowed Borlaug not one whit. He continued to be an agricultural ambassador up into his seventies, with his methods becoming widely acclaimed, especially outside of the U.S. It was in the 1980s, when Borlaug was semi-retired, that a notorious Japanese shipbuilding magnate was struck by images of starvation in sub-Saharan Africa. Ryoichi Sasakawa, chairman of the Japan Shipbuilding Industry Foundation, had experienced about as broad a career as one can imagine. His past included an association with the Japanese war machine and questionable business practices, along with international industrial success, and being a friend of Jimmy Carter, the first U.S. president to visit sub-Saharan Africa while in office. In the 1980s, Sasakawa began donating millions to the United Nations to help fight hunger in Africa. But as the years passed, he felt that not only was his money not enough, but that it might be going for weapons, not food. So, in 1985, he had his assistant call Borlaug and ask, "Why has there been no Green Revolution in sub-Saharan Africa?" The assistant asked Borlaug to bring his revolution to the last starving continent.

For more than three decades, Norman Borlaug had expended a huge amount of his time away from his wife and kids, traveling all over the world. For decades he had physically labored hours on end in the fields, putting wear on his body matching any laborer. If anyone had earned the reward of retiring, giving an occasional lecture, and bathing in the warm social status of sitting on some policy boards, it was Norman Borlaug. He knew little about Africa. Caught off guard, he told the assistant, "I'm too old to start learning now." The next day, Borlaug received a phone call from Sasakawa himself. The billionaire told him, "I am 13 years older than you are, Dr. Borlaug. We should have started sooner and didn't, so let's start tomorrow!"

Borlaug had never been one to turn away from a challenge to help people. And so, at age 72, he was called back into the battle to combat hunger. He organized a meeting of experts in Geneva, Switzerland, to evaluate the issue. Three conclusions emerged. First, the food crisis in Africa was the direct result of the political anarchy that had ruled much of the continent for decades. Second, as Borlaug asserted, "Lack of infrastructure is killing Africa." He explained that the United States had 13,000 miles of paved road for every million people, while Ethiopia had forty-one miles for each million people. How could there exist efficient markets with no way to transport the food? And finally, because wheat could only be grown in some parts of Africa, Africa would require additional food crops bred for better yields.

Jimmy Carter and Bill Foege, head of the newly-formed Carter Foundation, attended the Geneva Conference. There, Sasakawa turned his persuasive powers on the former president, and in late 1985 Global 2000, Inc. was created to address the starving continent. Carter and Sasakawa were co-chairmen, and Borlaug headed the agricultural initiative. Africa became Borlaug's new India.

The organization selected Ghana as its first country of focus in 1987, instituting the classic Borlaug-inspired agricultural initiatives. By 1991, Ghana was producing all its own food and a few years later actually began exporting food to other countries. One secret to the country's success was a new form of corn: Quality Protein Maize (QPM), a grain that is so nutritionally complete it could be used to wean infants off of breast milk. Unfortunately, the Ghana experiment has not been reproduced throughout Africa. The main challenges remain: a lack of infrastructure and warring governments, which turn millions of citizens into refugees and make agricultural production impossible.

Borlaug Today

Borlaug has remained on the world stage, preaching against complacency. As he warned in his Nobel speech: "The Green Revolution has won a temporary success in man's war against hunger and deprivation; it has given man a breathing space. If fully implemented, the revolution can provide sufficient food for sustenance during the next three decades. But the frightening power of human reproduction must be curbed; otherwise the success of the Green Revolution will be ephemeral only."

Consider: When Borlaug was born, the world's population stood at about 1.7 billion. When he won the Nobel Prize, it was at 3.7 billion. Today it stands at 6.5 billion, with an average growth rate of 1.1 percent per year. The population bomb did indeed explode during his lifetime. Borlaug asserts, "I am confident that the Earth can provide food for as many as ten billion people—six times the number who lived when I was born—if, and this is a big if, the world's societies support a steady stream of both conventional and biotechnology research, and political policymakers stay attuned to the needs of rural development."

Norman Borlaug, 94 years old when this chapter was written, continues to try. For more then 85 years he's been working to grow food, first as a farm boy, next as an agricultural scientist, then as a world agricultural diplomat, and finally, when he was 72 years old, in Africa. Despite his age, the man seems never to rest. "He's driven," says Hesser. "He obviously doesn't need the money; he spends hardly anything at all. But it's just that he wants to accomplish something significant while he's still here."

Lives Saved: Over 245 Million

Discovery: Green Revolution

Crucial Contributors

Norman Borlaug: Developed new strains of short stature, disease-resistant wheat, and trained scientists, farmers and governments in developing countries all over the world on how to grow the wheat using an integrated technological approach, including continuous breeding programs, fertilizers, pest and weed control, irrigation, and the implementation of profarmer government policies. Enabled countries like Mexico, Pakistan, and India to greatly improve nutrition for their populations and to become self-sufficient in their food production. His revolution then spread to China and much of the rest of the world.

**Norman Borlaug in
a Toluca, Mexico
wheat field.**
Bettmann Standard RM
Collection/Bettman/
CORBIS.

To produce 1999's cereal crop using pre-green revolution methods would have required an additional 3 million square miles to be farmed—about the size of the contiguous United States. At one time, Borlaug's wheat produced 23 percent of the world's calories.

"Don't try to discourage me, Ed. I know how much work is involved. Don't tell me what can't be done. Tell me what needs to be done—and let me do it. To hell with the extra work and strain. It's got to be done, and I believe I can do it."

—Norman Borlaug

"It would be helpful when you're working on these problems to develop a skin as thick as a rhino's hide, so you don't feel all the darts. Oh, there are lots of critics. If you don't do anything you'll never have critics."

—Norman Borlaug

When Richard Nixon visited China in 1972 on his critical détente mission Norman Borlaug accompanied him, underlying the promise the green revolution offered.

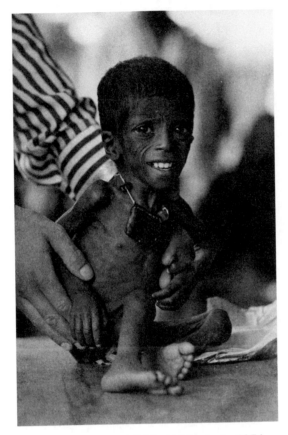

Starving Danakil child in Ethiopia in 1974.
Henri Bureau/Sygma/Corbis Collection/CORBIS.

A child that has mild malnutrition has twice the chance of dying from childhood diseases than a child who is well nourished. One with severe malnutrition has an eight-fold chance of dying.

Studies from Civil War soldiers showed that the taller a soldier was, the longer he lived. Childhood nutrition has a marked affect on height.

In 1980, just after the start of the Green Revolution, 47 percent of all children in the developing world had stunted growth. By 2000 the percent had declined to 33 percent.

In the Northern Hemisphere, people who are born in the months of October, November, and December live a half-year longer on average than those born in other months. This is thought to be because their mothers have better nutrition in the summer due to fresh produce.

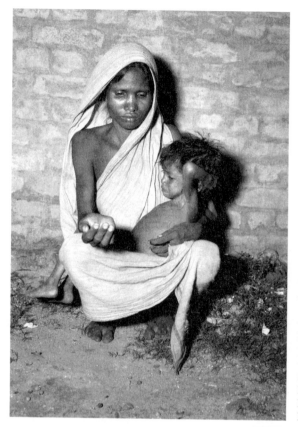

Starving mother and child in 1943 Calcutta. Before the Green Revolution, India had frequent famines.
Bettmann Standard RM Collection/ Bettmann/CORBIS.

"A person who has food has many problems. A person who has no food has only one problem."

—Chinese proverb

"You have to understand that Norman Borlaug has no ego. He's the world's greatest humanist. He cannot stand to see people suffer."

—Richard Zeyen, professor at University of Minnesota

"Dr. Borlaug's scientific achievements prevented mass starvation and death in South Asia and the Middle East. I have been particularly impressed by his work in Africa.... Dr. Borlaug is an American hero and a world icon."

—President George H. W. Bush

"It was fun to be there (in Asia after the Green Revolution) at times when younger children were regularly taller than their older siblings, indicating a dramatic improvement in both maternal and child nutrition."

—Thomas R. DeGregori, Professor of Economics, University of Houston

Chapter 6: John Enders—
Over 114 Million Lives Saved

The Father of Modern Vaccines

V ACCINATION BEGAN in the late 1700s when Edward Jenner inoculated a young boy against the smallpox virus, using a scab from someone infected with the much milder but related cowpox virus. To honor Jenner, Louis Pasteur later termed all such immunizing medicines as vaccines ("vacca" being Latin for cow), and the name stuck. However, even as late as 1950, there seemed to be a particular barrier preventing the development of viral vaccines. While the diseases caused by viruses were those for which the need for vaccines was greatest, there were no cures for any of them and, at best, doctors could treat the symptoms.

Then everything changed thanks to research led by a man you have probably never heard of. His laboratory pioneered the techniques that quickly led to dozens of vaccines, and he was personally instrumental in developing one of the most important of them. How important was it? Imagine a disease that everyone in the United States got, until a scientific advance reduced its occurrence to near zero. Of all the common human diseases, measles is the most contagious. Prior to the development of a vaccine in 1963, getting measles was a childhood rite of passage. Compare the two million yearly cases in the U.S. before a vaccination existed to the thirty-seven known cases in 2004, all of which were traceable to importation from other countries. Think about it: no rashes, no fevers, no complications from pneumonia or encephalitis, which can lead to death. Measles is a disease that is now gone, with no longer a dime spent on it beyond the cost of producing the vaccine.

A Member of the Lost Generation

John Enders was born on February 10, 1897, into a family of high-achievers. His grandfather had walked from town to town selling insurance, working his way up the ladder of the American dream until he became president of the Aetna Insurance Company. His father was president of Hartford National Bank. A frequent family visitor during Enders' childhood wore a spotless white suit and was better known by his pseudonym Mark Twain. In school, the young Enders had trouble with math and physics, and later said that during those years he "preferred in the main certain of the so-called humanities—Latin, French, German, and English literature, although biological subjects always proved highly attractive." At 17, he was already concerned about accomplishing something in life, writing:

When I am old
And spindle shanks and sunken chaps
Betoken the coming of the fearful
    Last Event,
What memories of youth will come to
    steel my heart
Against the fear of meeting with my
    Creator?
Will then my heart be strong with
    thoughts of Deeds well done?
Of struggles, great achievements, the
    joy of children?
Of weeping, sorrow, pain—a life lived
    to its full?
Or shall I gaze with sorrow and despair
Back upon the assemblage of empty,
    wasted years
And curse a life of dull and brutish
    Indolence?

He matriculated to Yale University, but interrupted his education to enlist in the Naval Reserve to serve in World War I. There he learned to pilot the fragile, rickety bi-planes of the time. Air flight was only 15 years old, and these daring pilots sat in cold, open cockpits. Enders was not sent to combat, but became a flight instructor in Pensacola, Florida. If he had any notions of being a daredevil himself, his experiences flying completely cured him. He decided he had used up all his luck in the air, and forever after was extremely reluctant to fly.

After graduating from Yale, Enders set out to uphold the family tradition of business. Brokering real estate was an easy entry, but he had trouble seeing the value of convincing someone to buy something they already wanted, and never could gain any enthusiasm for it. His twenties were well described by his contemporary Ernest Hemingway, who wrote about the Lost Generation of young adults too disillusioned after World War I to find purpose in life. After four fruitless years in business, Enders searched for something to interest him and, finding nothing, decided to go back to school. Choosing Harvard this time, he received a masters degree in English, then decided to pursue a doctorate in philology, the study of language as used in literature. His first published academic paper was on the use of the word—*Naometria*—in Ben Johnson's play *The Staple of News*. It was nothing he could get enthused about. Writing a friend he said, "I mouth the strange syllables of ten forgotten languages, letting my spirits fall, my youth pass. If this mood lasts, I shall, by Heaven, throw it all to the four winds and go forth into the world like Faust, even if I have to bear his penalty."

While lost in purpose, Enders' life was not miserable. He lived with some medical students in Mrs. Patch's comfortable boarding house. There he met an animated Australian, Hugh Ward, whose daily description of his doctoral work in microbiology piqued Enders' interest. "We soon became friends," Enders wrote, "and thus I fell into the habit of going to the laboratory with him in the evening and watching him work. I became increasingly fascinated by the subject—which manifestly gave him so much pleasure and about which he talked with such enthusiasm." Ward also introduced him to Hans Zinsser, a charismatic professor whom Enders described as, "a man of superlative energy. Literature, politics, history, and science—all he discussed with spontaneity and without self-consciousness. Everything was illuminated by an apt allusion drawn from the most diverse sources, or by a witty tale. Voltaire seemed just around the corner, and Laurence Sterne upon the stair.... Under such influences, the laboratory became much more than a place just to work and teach; it became a way of life."

Together, Ward and Zinsser transformed Enders' life, convincing him that his intellect was better suited to science than philology. So in 1927, at the age of 30, considered late even in his day, Enders made a 180 degree turn in his studies. He said, "This antipodal revolution of my studies has been of large value in helping me to obtain that Pisgah [the mountain which Moses ascended to see the Promised Land] sight of things and people that perhaps is the ultimate aim of my apparently inconsistent, faltering and obscure action." He entered the doctoral program in bacteriology at Harvard and, with an ease that seemed to surprise even himself, took to the lab work. That same year he married Sarah Frances Bennett.

It was rare to earn a doctorate under Zinsser, a demanding taskmaster, but Enders earned his in 1930, with a thesis on anaphylactic shock caused by carbohydrates extracted from tubercular bacteria. He became an instructor and worked in the Harvard labs. As a teacher, he was noted for his personal magnetism in small-group and one-on-one settings, including in the lab, but he seemed less suited to giving lectures.

"Delivery was another matter," said a later colleague, Samuel Katz. "After adjusting the microphone around his neck and checking that it was live, he would extract a large handkerchief from his vest pocket and blow his nose with significant amplification by the sound system. We were never certain whether that was a nervous habit or meant deliberately to arouse the students' attention or laughter, but it usually did both. He extracted a large gold pocket watch from another compartment of his vest and placed it on the lectern to assure punctuality in his presentation. The lecture was articulate, but dry, and rarely overwhelmed the audience. His own disciples sat squirming in the back row seats, hoping the students would perceive the gems they were receiving. After the lecture, we joined him for the short walk back to the laboratory in the old Carnegie Building of Chidren's Hospital, assuring him of how fine his presentation had been, even though both he and we suspected that few of the students appreciated what had been offered."

Perhaps because he had found his intellectual home in Zinsser's lab, Enders showed little of the academic ambition typical of Ph.D.'s in their 30s. It was five years before he became even an assistant professor and another seven years before he became an associate professor. It would be thirteen years before he could employ a personal technician. Research, as opposed to advancing his career, was his main interest, so he didn't complain. When, in 1937, Harvard's experimental kittens came down with feline distemper, Enders and a young epidemiologist, William Hammon, began a careful study of the disease. They found it was caused by a virus and worked on developing a vaccine. This and other studies turned Enders' interest toward viruses, where he made his first major breakthrough—"the development of serologic techniques for the detection of antibodies to the mumps virus."

Viruses

Viruses are very different than bacteria, the other type of agent that causes most disease. They are so tiny that they were not even seen until the advent of the electron microscope in the 1930s, when Enders was beginning his research. Made up of genetic material—either DNA or RNA strands—and encapsulated in a protective protein shell, viruses thrive in bacteria, fungi, plant, and animal cells, co-opting the cells' genetic machinery to acquire energy to reproduce. It is arguable whether they are even life forms, but they definitely impact humans. When they parasitically infect our bodies, they cause diseases such as the common cold, flu, hepatitis, herpes, HIV, SARS, and ebola, just to name a few.

Viruses' exceedingly small size (hundreds or thousands of viruses can exist inside a single human cell) makes them devilishly difficult to study. At the time that Enders began to take an interest in them, most viruses could only be grown in vivo (Latin for "in the living"), which meant in living animals. Animal experiments are expensive, slow, and awkward; plus, they often require the sacrificing of live animals when a scientist wants to observe the effect of a particular trial.

What scientists needed were techniques for studying viruses in vitro (in test tubes), which presented several challenges. First, a given virus will only grow in certain cells from certain animals. Second, the cells must be healthy when infected with a virus, and a test tube is not a naturally healthy setting for a living cell. Third, the cells must remain healthy, which means they must be provided with nutritious food. Fourth, the cells must remain uncontaminated, but in a test tube, bacteria, other viruses, and fungi can easily contaminate and kill cells. Finally, if a virus can be induced to multiply, it often kills the very cells into which it has been cultured, releasing viruses into the liquid remnants. The only way to maintain those viruses is to transfer them to a new tissue culture, called a subculture. All of this has to be repeated over and over again. In the 1940s, growing viruses in test tubes in sufficient quantity to experiment with was a major barrier to virology research.

The Iron Lung

The early 1940s greatly upset Enders' idyllic lab life. In September of 1940, Hans Zinsser died, not only costing Enders his cherished mentor, but also burdening him with the lab's bureaucratic administrative duties, which were not his forte. Then World War II began, resulting in most of the young doctoral candidates leaving for military service and thus ending many lines of research. Tragedy struck again in 1943, when Enders' wife Sarah suddenly died from acute myocarditis, an inflammation of the heart.

The forties were particularly tragic for another reason—polio epidemics. And it was out of polio research that the techniques leading to the development of the measles vaccine emerged. Karl Landsteiner published the basic facts about polio in 1909. It was a virus, and monkeys were the experimental animals of choice since it also infects them. There was, and still is, no cure.

When polio struck, terror followed, and if you had children, panic was completely understandable. Readers over the age of 60 remember it well because it lurked in the dark shadows of their nightmares when they were children. Unlike anything parents face now, polio came like clockwork when the weather warmed and left with the first chill. Parents examined their children weekly during the summer, pumping their legs to look for the telltale weakness, touching their foreheads to check their temperatures. Most of the time the epidemic was mild and most of the victims, largely children, recovered without permanent damage. Every few years, however, it struck with particular viciousness. In 1916, New York City tallied 9,300 cases, with 2,200 children dying. Half the survivors were paralyzed.

A decade later, scientists developed a respirator, technically known as a "negative pressure ventilator," that kept alive children who could not breathe because of paralysis of the diaphragm. Most people called them "iron lungs," although the name sounds more like a medieval torture chamber than a lifesaver. Invented at Harvard in the late 1920s, and actually made of steel, the iron lung consisted of a drum that sealed the body from the chest down, leaving the head and shoulders free. Pumps increased and decreased air pressure in the chamber. When the pressure in the drum fell below the level of the pressure in the lungs, the lungs expanded and sucked in air; when the pressure was greater outside the lungs, the lungs were squeezed to exhale. Hospital wards were filled with them.

Polio struck the wrong person in 1921, the future President Franklin Delano Roosevelt. FDR was grown when he was diagnosed and, being from a wealthy family, turned many resources toward finding a cure, eventually founding the National Foundation for Infantile Paralysis. Promoted by Basil O'Conner and Carl Byior, the National Foundation became the largest voluntary health organization in the country and a model for modern private charities. When FDR became president, the comedian Eddie Cantor became its spokesman and suggested people mail dimes to the White House, directly to FDR. Cantor also thought up the name March of Dimes. He went on the radio to tout his plan, joined by the Lone Ranger, Jack Benny, Bing Crosby, and Rudy Vallée. On an average day, the White House got 5,000 letters. The day after the Cantor show, 30,000 arrived. The next day the count rose to 50,000, and on the third day 150,000. When volunteers had cleared out the hallways and desktops, they counted 2,680,000 dimes.

Almost every authority on the disease in America eventually either worked directly for the foundation or was funded by it. Its budget dwarfed that of the National Institutes of Health, which had only recently been chartered with a tiny budget. It is no wonder that Jonas Salk became the most famous medical scientist in the nation's history. Terrifying the public with a constant flow of news items, getting people personally involved through donations, tugging on people's heart strings with crippled poster children, O'Connor and Byior practically invented modern public relations.

A Poor Polio Paradigm

Prior to the National Foundation for Infantile Paralysis, the Rockefeller Institute for Medical Research funded most medical research. Founded in 1901 by John D. Rockefeller, under attack at the time for his company, Standard Oil's monopolistic practices, the Rockefeller Institute produced impressive research, decades later financing most of Howard Florey's work on penicillin and Norman Borlaug's Green Revolution. The institute's first director, whose tenure lasted more than three decades, was Simon Flexner. He had discovered a treatment for a type of meningitis before the discovery of antibiotics.

Researchers under Flexner made several mistaken conclusions concerning polio. One led to a misconception that, propagated by Flexner, held back polio research for over a decade. Flexner's researchers proved that polio was neurotropic, meaning it multiplied in nerve tissue, which Flexner concluded as meaning polio multiplied only in nerve tissue. Because the highly respected Flexner said so, it was the end of research in that direction.

This idea fit into the common beliefs of the time. Alice Huang, a professor of microbiology and molecular genetics at Harvard, says, "Many scientists at that time persisted with the notion that growth of poliovirus—or any other infectious agent, for that matter—could take place only in the target cells. Of course, all these efforts failed to produce any substantial amount of poliovirus."

Indeed, in the years prior to World War II, some researchers attempted, unsuccessfully, to make a vaccine from the small amount of virus that could be obtained from monkey spinal cords. It became clear that vaccine creation would require both an accurate paradigm and a way to dramatically multiply the poliovirus, something which had proven very difficult to accomplish in nerve cells.

The Gentleman Scholar from Connecticut Gets His Own Lab

After the war, Enders was asked to establish a research laboratory on infectious diseases at Boston Children's Hospital, which was located just across the street from Harvard Medical School and had various ties to Harvard. Today, that may sound like an odd place to do research, but in the 1940s it was ideal. Researchers were drawn to pediatrics because children were both the battleground and the living laboratories of infectious microorganisms like polio and measles. In 1940, Enders had worked with Thomas Weller, a young medical student, on an experiment that had revolutionary implications. By using a roller tube apparatus, they had kept cowpox virus (*vaccinia*) growing for nine weeks in a test tube. The previous viral life span in tissue culture had usually been measured in hours or days, not weeks. Now that Weller was back from military duty, Enders brought him on to be the assistant director and to take care of administrative duties.

The hospital allocated four rooms on the second floor of the Carnegie Building, a three-story building next to Harvard's coal-fired power plant. First it needed a thorough cleaning—oily soot had come through the loose window frames and coated the fixtures. Then, two rooms were turned into labs; one room was the glass washroom, and the final room was Enders' office. On the wall Enders hung pictures of Edward Jenner and Louis Pasteur, and would later add pictures of researchers who worked in the lab, each with a story Enders loved to tell. Animals were kept a block away in another building. Enders and Weller furnished the lab with the standard equipment, such as an autoclave to sterilize equipment and chemical hoods to work under, along with some more novel items. Embryonated hens' eggs had recently been introduced as media in which to grow viruses, so they purchased an egg incubator and visualizing lights to shine into the eggs. Turning the eggs was a daily chore, including weekends, but one that had a side benefit. Not all of any batch of eggs were fertile, so the young researchers received a few free eggs from time to time to take home.

Early in 1948, they added Weller's college roommate Frederick Robbins to the team. Robbins and Wellers were both brilliant doctors, but very different in personality. Robbins was relaxed and reflective, Weller dynamic and animated.

A technician in the lab, Alice Northrop, says, "We all liked each other and admired Dr. Enders, whom we called 'the Boss.' Because Dr. Enders did not wish to eat at the hospital, which Tom Weller always did, the rest of us had lunch at the big table in the large lab adjacent to Dr. Enders's office. Jeanette, Carol, and I made the lunch, being careful not to have anything containing egg, as Dr. Enders was allergic to eggs. "Does it have egg in it, Alice?" he would say in a worried voice. Dr. Enders had unusually broad interests and enjoyed discussions in the lab at lunch about all sorts of things as well as work: politics, literature, the arts, even exciting sailing adventures." Enders' sly sense of humor often came out, whereupon he would give a half-grin and his eyes would twinkle.

Enders never seemed in a hurry in his life, and by the time they had the lab up and running, he was 50 years old, slightly stoop-shouldered, and tweedy. He would arrive at the lab at nine or ten each morning. His first priority was looking at new findings, so he made the rounds to all the experiments, moving slowly, often with a pipe in his mouth. Northrop relates, "He would often say, 'I've been thinking this morning, Fred,' and then describe an idea for an experiment." He was never curt or dictatorial, and enjoyed discussing ideas the other researchers had. David Tyrrell, a virologist who knew Enders, said, "His relations with the staff were such that they were able, and indeed encouraged, to produce suggestions for the next experiment or a new idea to solve a problem. This would be discussed from all angles and quite often the chief would put forward a different idea, but in such a way that the trainee was glad to take it up rather than feeling resentful that he had been forced to abandon his own idea and accept that of 'the Boss.'"

The aim of the lab, which many thought futile, was to study numerous viruses in tissue cultures in test tubes. Enders had already worked with viruses such as herpes simplex, influenza, measles, and mumps, and wanted the new lab to expand such research. There were numerous experiments going on at any one time. Enders summed up the time he spent in the 1930s: "In these experiments the tissue culture method was employed with uncertain results. But the conviction was gained that it represented a basic tool for the study of viruses of which the possible applications were almost unlimited."

Almost Without Conscious Effort

Weller was full of ideas and set about trying to isolate the varicella virus, the cause of chicken pox and shingles, using embryonated hens' eggs. The experiment failed. Early in 1948, he decided to construct an experiment that might work in a way similar to their 1940 experiment, still trying to isolate the varicella virus. Enders suggested he try it with the mumps virus instead. Although Enders had failed earlier to grow mumps in tissue culture, he thought that the accumulation of new techniques might now allow it to work.

Weller's idea was to start with the Maitland technique, whereby tissue is placed in a flask along with a nutrient solution, then inoculated with a virus. He inoculated chicken eggs' amniotic membranes (the fluid sac that surrounds and protects the embryo) with the mumps virus, and placed it in a flask with a nutrient mixture of balanced salt solution and ox serum ultrafiltrate. With the Maitland technique, after a short time the fluid containing the virus is drawn off and put into another flask containing fresh tissue, making a subculture. Weller's idea was to modify this technique by leaving the tissue in the flask and replacing the nutrient fluid every four days. This allowed the tissue to remain alive for a longer period of time, giving the virus more time to grow in each cell. To see if the experiment had succeeded, Weller measured the concentration of hemagglutinins—a class of protein that viruses use to attack host cells—in fluid they drew off.

The experiment worked. The tissue did not quickly die, and the virus continued to multiply and grow. Enders suggested they do the same experiment using an influenza virus. They isolated it as well, and found they could keep tissue growing for thirty days or longer.

The Growth of Poliovirus in a Flask

Next, Weller prepared to use this technique to try to isolate the chicken pox virus, as he had originally intended. Since the varicella virus was only known to grow in human cells, he needed to obtain human tissue. At 8:30 on a March morning, he went to the Boston Lying-in Hospital (a maternity hospital), where he obtained aborted human embryonic tissue. That afternoon he prepared cultures of skin and muscle tissue, minced them, and placed them into flasks. He inoculated four of them with cultures of throat washings from a child who had chicken pox. Another four flasks were not inoculated and were to act as controls.

He had enough skin and muscle culture left for four more flasks. There were three strains of poliovirus known at that time. The primary human pathogen was the Brunhilde polio strain, which had previously only been grown in humans and monkeys. The Leon strain was very rare, but significantly a third strain, the Lansing strain, had been found to grow in mice. The lab had cultures of it and Weller put it in the four remaining flasks. He recalls, "It was almost as an afterthought: I was focusing on growing the varicella virus, not the poliovirus."

Four days later, Weller changed the nutrient fluid in all twelve flasks, and changed it again on April 7, whereupon he took samples to test for growth of the chicken pox virus. He was disappointed: There was no evidence that the chicken pox virus was growing in the flasks. He also took fluids of the cultures he had inoculated with poliovirus and injected them into the brains of five mice. On April 14, one mouse became paralyzed and died and, as the days passed, three more mice died.

Enders kept several novelty hats on a shelf in his office; when exciting experiments were to be examined, he would put one on. If ever there was a time to don such a hat, this was it, for the experiment demonstrated something new. Never before had there been evidence that poliovirus could be grown in non-nerve cells.

In their Nobel lecture, Enders and his co-winners said, "...from time to time we had considered the mounting evidence...in favor of the possibility that these agents [of polio] might not be strict neurotropes.... Such ideas were in our minds when the decision was taken to use a mixture of human embryonic skin and muscle tissue in suspended cell cultures in the hope that the virus of varicella might multiply in the cells of its natural host.... While close at hand in the storage cabinet was the Lansing strain of poliomyelitis virus. Thereupon, it suddenly occurred to us that everything had been prepared almost without conscious effort on our part for a new attempt to cultivate the agent in extraneural tissue."

Enders now focused the lab's efforts on the poliovirus. Tyrrell recalled that Enders "favored doing small-scale experiments and reviewing the results of each as they came out to define another simple question and plan the next experiment to answer it." Enders had Robbins repeat the experiment, using the intestinal tissue obtained from an autopsy of an infant who had been born prematurely and had died. This experiment, too, was successful. After running tests to rule out other viruses that might have contaminated their cultures, they tested it on two monkeys. Both monkeys became paralyzed. Other agents could cause paralysis, however, so they autopsied the monkeys. The autopsy of the monkeys' spinal cords showed the telltale signs of poliovirus—their large nerve cells were shrunken and their nuclei were knotted.

Enders and his colleagues knew well they were dealing with a dangerous pathogen, and they had already developed safety procedures for handling viral specimens. Weller didn't even want his wife, Kay, to visit the lab. To be extra cautious, they put the two sacrificed monkeys into heavy bags, lugged them back to the lab, and put them into the autoclave; doing so would kill any virus that remained alive. The whole floor stank.

Further experiments showed they could grow the other two known types of polio, the Brunhilde and Leon strains in the same manner. Over the next year Enders could don a celebratory hat often. He and his colleagues demonstrated that the virus would grow in human embryonic tissues from the intestine, liver, kidney, adrenal, brain, heart, spleen, and lung. The Flexner polio paradigm was completely overturned.

Cytopathic Effect

Enders ' work with the mumps virus had shown that fluids infected with the virus visibly affected red blood cells, which could be observed under a microscope. This provided an important lab test that could be used to tell if a fluid was really infected by the mumps virus. If he and his colleagues could find a similar test for the polio virus, showing that it infected and harmed tissue cells, it would be much easier than autopsying monkeys.

One day, peering into the microscope, Enders saw a poliovirus-infected cell explode. "Cytopathogenicity," he exclaimed, inventing a word. The word, truncated to cytopathic, later became common in virology, used to describe the damage viruses cause to cells. He and his team could observe the pathological effect of the poliovirus on the cultured tissue cells sixteen to thirty-two days after being inoculated.

Next they implemented a chemical marker identifying the growth of the virus in infected tissue by adding phenol red, a dye. The poliovirus affected the cells' metabolism, such that the cells produced less acid than those not infected. In uninfected control cells, the dye changed color to yellow as the cells acidified. In infected cells, it remained red. Thus they had a ready test for the presence of polio in human tissues. The researchers also discovered a positive correlation between the acidity of the cultures inoculated with poliovirus and the amount of time since inoculation. Such precision allowed them to measure the infectivity of the virus, as well as to determine the minimum dose needed to infect various types of tissue.

The team shortened its ability to recognize the virus to eight days using another technique, and found a way to develop a test tube test to determine the type of poliovirus and their respective antibodies in human or animal blood serum. These techniques would all become common testing techniques in virology, although not all viruses produce visible changes under the microscope, nor do they produce the same chemical changes as poliovirus.

The Roller Tube Apparatus

Roller tube tissue cultures had been perfected by George Gey in 1933, and Enders had been one of the first to use them with viruses. Rather than growing animal tissue in flasks, researchers, using roller tubes, plant tissue cells in plasma clots on the sides of test tubes. Viruses can then be inoculated into this tissue. A nutritious fluid consisting of inorganic salts, serum from animals or humans, and an extract from embryonic tissue is supplied to keep the tissue alive. The capped test tubes are then kept almost horizontal and heated in an incubator to 37 degrees Centigrade (99°F.), which is approximately the temperature of the human body. The test tube is slowly turned at a rate of eight to ten times an hour, exposing the tissue to both the liquid nutrient and air. The goal is to more accurately mimic conditions inside the human body, where tissue is exposed to an active environment of nutrients and waste product removal, rather than the static environment of a motionless test tube or flask.

A big advantage of using the roller tube technique comes when viewing the cultures under a microscope. Tissue from flasks has to be chopped, and then prepared as slides for the microscope, while the outgrowth of cells into the plasma clot of a roller tube can be repeatedly examined in the living state as the cells grow and the virus replicates.

A handy carpenter made the researchers a roller tube apparatus with a round wooden holder of about 100 test tubes. When they used the roller tubes, they found that the virus grew more rapidly and to higher titres (concentration) than in the suspended cultures in the flasks. This cut the time necessary to observe the growing virus to four days.

Leaving Monkeys at Play
by Adding Antibiotics

Until that point, scientists who had wanted to isolate and type polio virus cultures could only do so by injecting the virus directly into the brains of living monkeys, letting the virus infect the brain tissue, killing the monkeys, and testing samples for virus. In the end, more than 100,000 monkeys were sacrificed in the development of an as-yet only dreamed of polio vaccine. Through their constant push for greater experimental efficiency, however, Enders and his team found a way to spare future monkeys by successfully isolating poliovirus from the monkeys' feces. Robbins performed much of this research, adding penicillin and streptomycin—two newly discovered antibiotics—to kill off bacteria. He then centrifuged the purified samples and inoculated tissue cultures with them, which quickly showed the telltale signs of polio infection. They found that they could often isolate and type virus from human feces within forty-eight hours, as well. Enders recognized the importance advances in other fields could have on his own research, stating in the team's Nobel Prize acceptance speech, "At this point it seems appropriate to remark that the discovery of the antibiotics has, as in so many other areas, worked a revolution in the field of tissue culture. Through the use of these substances it is now not only possible to apply tissue cultures to the routine isolation of viruses from materials heavily contaminated with microorganisms, but it has become feasible to use them under conditions and in numbers which in the past would have been quite unthinkable. Here then we have another example of how one discovery leads to many others, often of quite a different nature."

When Robbins found something exceptional, he wrote the results in his laboratory notebook in red ink. The red ink flowed. Experiments demonstrated that many children carry lots of viruses similar to polio, often with no symptoms. Within two years, the team had isolated and typed thirteen strains of poliovirus. They also discovered the first of a whole group of viruses that became known as ECHO viruses (enteric cytopathic human orphan viruses), cousins to polio.

During this time, the Enders team were writing papers and giving frequent presentations of their findings, so many researchers would visit. Jonas Salk was one. Katz says, "The door was always open to visitors from throughout the world. Few scientists left such a visit unaccompanied by carefully packaged boxes containing samples of virus, cells, sera, reagents, or other ingredients, to ensure the ready progress of their own experiments back at home. The data in laboratory notebooks were shared with visitors who sought specific information on experiments recently completed or still underway.... His philosophy was that the more people working on a problem, the sooner a solution would be found." Salk offered Robbins a job, which Robbins turned down.

The Holy Grail—Attenuation

Weller continued to work on propagating the poliovirus. He grew the Lansing virus through twenty-three passages (a passage occurring whenever he changed the tissues in which the virus was growing), over 331 days. He grew the Brunhilde strain—the dominant cause of polio in humans—through fifteen passages over a period of 267 days. Periodically, the researchers tested it. They found that virulence had declined after three passages. As the number of passages increased, the virus's virulence decreased well over 100,000-fold.

The weakening of a virus through serial passage was a method developed by Louis Pasteur in the 1870s, arising from his work with chicken cholera, anthrax, and rabies. Serial passage exploits evolution through the principle of natural selection. Multiplying viruses will mutate in many different ways, and the ones that survive are those that are best adapted to living in whatever tissue they happen to be cultured in. A virus taken from a chicken and then cultured and subcultured several times in human tissue will lose most of its ability to infect the chicken. Likewise, a virus that is devastating to humans can be passed serially through various non-human tissues and end up vastly weakened—known in virology as attenuation—yet retain enough of its original protein coat signature that the antibodies a human immune system produces to fight the attenuated virus in the vaccine will also fight the original virus. This "playing with evolution" is time-consuming and arduous. A virus has to be grown over and over again in different media until it mutates into the desired, weaker form.

The conclusion was spelled out in their Nobel Prize lecture: "From these observations we concluded that, as with other viral agents, the virulence of poliomyelitis virus is not a fixed attribute but on the contrary may readily be altered under appropriate conditions." In less understated words—a vaccine was possible!

A Revolution in Molecular Virology

In the 1870s, Robert Koch invented bacteriology by assembling its basic tools. He began by trying to grow bacteria using potato slices, but soon moved on to gelatin broths. These worked well at cold temperatures, but not at hot ones. Others suggested using a type of Japanese seaweed called *agar-agar* which could be made into a waxlike medium. Richard Petri added the dish. Using a petri dish and agar, and adding a sugar-rich substance as a nutrient, bacteriologists found they could grow, identify, and study all sorts of bacteria. Thus microbiology was born.

Enders and his colleagues sparked a comparable revolution with the development of their tissue culture techniques. Viruses had been grown in tissue cultures prior to Enders, but only sporadically and with difficulty. As a result, most researchers used live animals as the medium in which to grow viruses. Imagine you are a virologist and every time you have an idea, you have to inject an animal with a virus, wait for it to incubate, kill the animal, and then cut it up and examine the tissue; and if in doing so you get another flash of insight, you have to do it all again. Enders' team devised methods that provided virologists with the ability to grow a limitless supply of virus in numerous tissues, since they had shown that they could achieve viral growth on the order of 10^{15} (one quadrillion!) times the amount of their initial experiments. They demonstrated the value of looking for cytopathic cell damage due to the invasion of viruses. Moreover, their methods drastically shortened the time required for experiments, allowing for the testing of virtually endless ideas. Soon their methods were used in laboratories all over the world, leading to numerous major discoveries.

When Enders began his work, only thirteen human disease-causing viruses had been isolated. By 1961, fifty-eight more human viral pathogens had been cultured, plus another 300 viruses that infect animals. Finally, Enders' team demonstrated all the techniques necessary for each step that needed to be taken in the creation of a vaccine, information which led to a cascade of vaccines for a plethora of viral diseases.

Romance in the Lab

The lab began as a group of diverse researchers, but grew, literally, into a family. Shortly after Robbins joined the lab, he fell in love with Alice Northrop, and soon they married. Enders followed suit, marrying Carolyn Keane in 1951. Robbins writes that Carolyn's "cheerful disposition, facility in social intercourse, and great energy proved to be of enormous support." Fred Robbins thought so much of Enders that he suggested to Alice that they name their second daughter after him. She agreed, and the name Louise Enders Robbins was entered upon the birth certificate. Summers often found the Enders at their waterfront house in Connecticut, where John liked to go power boating on Long Island Sound, often fishing for striped bass.

A well-regarded tradition was the family Christmas party. Enders donned a crimson smoking jacket and plied his guests with drinks. Katz recalls, "One delightful tradition involved the drawing of names from a hat, a few weeks before the holiday, for the giving of Christmas presents. More attention was focused on the requisite poem which accompanied the gift, and which was read aloud at the time of presentation, than on the gift itself. Some of the efforts were indeed remarkable, but none matched those which Enders himself composed, and it was a true joy to be the recipient of his poem and gift. The evening ended with the Chief at the piano and his guests gathered round him, his 'family' singing Christmas Carols."

Salk and Sabin

A vaccine is a preventative medicine that jump-starts the immune system into producing antibodies to attack a specific disease, usually before the immunized person contracts the disease itself. Edward Jenner was fortunate that naturally occurring cowpox acted as a vaccine for a similar, although much more serious, disease—smallpox. It has been difficult to find such relationships for other diseases. Generally, scientists must try to manipulate pathogens in the laboratory to make a related, but asymptomatic, form of the pathogen that will act as a vaccine.

In the 1940s, there were vaccines for three childhood diseases—diphtheria, tetanus, and whooping cough (pertussis)—all caused by bacteria. For diphtheria and tetanus, researchers discovered how to make toxoid vaccines, which are based on purified and deactivated forms of the disease-causing toxins these bacteria produce. Whooping cough was a killed vaccine: the bacteria that caused it were killed, but retained the protein coat signature that allows our immune system to identify them. Before Enders, the only preventative viral vaccine was for yellow fever. Using the serial passage technique, researchers had grown the virus over and over again in different tissues until it had evolved into a much weaker pathogen. Each of these types of vaccines were shown to trigger a protective immune reaction in humans.

Interestingly, the Enders lab did not try to develop the polio vaccine, even though they had the knowledge and could have easily secured the funding. "Enders, in his thoughtful way," Robbins said, "felt that this was not the kind of work our laboratory was best suited for." So the team stepped back and passed the task to Jonas Salk and Albert Sabin, who already had big labs dedicated to vaccine development. Both had pretty much stalled before the breakthroughs made by Enders' team. Suddenly, with a way to obtain an endless supply of viruses, they made quick progress. Salk himself said, "Dr. Enders pitched a very long forward pass, and I happened to be in the right place to receive it."

In 1952, Salk developed a vaccine that contained chemically killed virus from all three types of polio. In what was then the largest public medical experiment in history, 200,000 children were injected with the Salk vaccine, and the vaccine proved almost 80 percent effective at preventing infection with polio. It came at a propitious time, for in

1952 polio again flared up. Cars lined up at the entrance of Boston Children's Hospital, with parents rushing children into the emergency room, where doctors performed triage. The Salk vaccine was licensed for use in 1955, and within the next decade immunization made polio epidemics a thing of the past in most of the developed world.

In 1957, Albert Sabin developed a vaccine of weakened, or attenuated, virus, which he insisted was safer and more effective than the Salk killed vaccine, and which could be taken orally in delicious sugar cubes. Because it was oral, Sabin's vaccine had the additional advantage of causing the body to produce antibodies in the intestine as well as in the bloodstream; therefore, unlike the killed vaccine, it prevented immunized people's intestines from serving as reservoirs for poliovirus. Nor did it require the booster shots that Salk's vaccine required. It was Sabin's live vaccine that became the standard vaccine used in the United States in 1961. Endorsed by the World Health Organization, it eventually was used throughout most of the world.

The Nobel Prize

With the work on polio finished, Robbins left in 1952 for a position as Professor of Pediatrics at Western Reserve University in Cleveland, and in 1954 Weller made the short move back to Harvard to become head of the Tropical Public Health Department. His virology work later isolated the viruses that cause chickenpox and rubella, among others.

In the fall of 1954, Enders, Weller, and Robbins were notified that they had received the Nobel Prize in Physiology or Medicine "for their discovery of the ability of poliomyelitis to grow in cultures of various types of tissues." While Salk and Sabin received most of the public recognition for developing the polio vaccines, it is noteworthy whom the Nobel Foundation chose to honor. Weller and his wife flew across the Atlantic, but the Robbins and Enders took the *Queen Elizabeth* across and, much to Robbins regret, braved the stormy North Sea on to Sweden. Everyone got seasick except Enders. It was the same year Linus Pauling won the prize for chemistry and Ernest Hemingway for literature.

Two years later, Harvard granted Enders a full professorship, after his twenty-six years of productive research. Meanwhile, he had already begun applying the new tissue culture techniques to another virus.

Measles—Astonishingly Communicable

Polio, even at the height of epidemics, did not strike a huge percentage of the world's population. However, the attention the panic-stricken U.S paid to the decades-long effort to develop a polio vaccine gave a huge boost to research into vaccines for many other diseases, one of which has saved a quantity of lives that is truly heroic.

Measles has been with us for thousands of years, and was described by Persian doctor Rhazes around 900 AD as "the safer form of smallpox." It was not until the 1600s that the two diseases were recognized as distinct, and efforts to inoculate against measles—similar to what was done to prevent smallpox—began in the next century, albeit without success. Measles usually strikes in waves of two to four years and is astonishingly communicable—even the slightest contact will pass it on. It remains incurable to this day, although secondary infections can be treated with antibiotics. As pediatrician Louis Z. Cooper described it at that time: "Before a measles vaccine was available, more than nine out of ten children caught measles, half of them before they were five. Measles starts with a high fever, a runny nose, a cough, and a sore throat. After five days, a blotchy red rash develops. Usually the disease runs its course in another week, but sometimes there are complications. One measles patient in every six develops pneumonia or a serious ear infection. One in every thousand gets measles encephalitis, an inflammation of the brain that can cause paralysis, mental retardation, and even death. Far from being a harmless childhood disease, measles kills more children than any other acute infectious illness."

Measles was a common childhood disease in America into the 1960s, although not a big killer, thanks to our country's good nutrition and medical care. But worldwide it was devastating. The World Health Organization estimates that when Enders began his work, 106 million people got measles each year, and more than 6 million, mostly children, died.

An Assault on Measles

Measles had been a particularly difficult virus to work with because there was no good animal host on which to experiment. When Enders began researching measles in 1953, Asian monkeys were the only nonhuman species that had been found to be susceptible to the virus. Unfortunately, sometimes a virus would infect them with measles and sometimes it would not.

Enders' new tissue culture techniques did not do away with all animal experimentation; monkeys were still needed to test vaccines upon. Without a test animal, and with the tissue culture technique still being relatively new, little progress had been made with the measles virus. Virtually everything needed to be done: isolate the virus, accurately determine how it infects cells, and learn about its basic properties as a pathogen, such as whether immunity could be expressed. Only then could an effort to attenuate it into a vaccine begin.

Samuel Katz worked with Enders during the measles research and knew him well. "The superficial Enders guise was that of a round-shouldered, overburdened, sometimes meek, pedantic scientist—a caricature he deliberately fostered and exploited effectively as a shield against the unwelcome intrusion of assignments to distracting committees or administrative chores. In fact, he was a strong, competitive, thoroughly contemporary, artful academician who conserved his energies for those challenges he judged worthy." In order to maintain personal interactions in the lab Enders, only accepted four or five fellows and a couple of faculty associates at any one time. Katz says Enders was "firmly convinced that the size or magnificence of laboratory surroundings and equipment had little relationship to productivity or success—if anything, an inverse relationship."

Six scientists in Enders' lab played significant roles for varying lengths of time in the assault on measles: Samuel Katz, Thomas Peebles, Kevin McCarthy, Anna Mitus, Milan Milovanovic, and Ann Holloway. "Nothing was too menial for a fellow," Katz recalls. "In order to learn thoroughly every detail of cell culture and virology techniques, the fellows shared responsibility with the technicians for the preparation of common pools of media, reagents and cell cultures." One typical task they all collaborated on was the preparation of human amnion cells. Every few weeks, they obtained placentas from Boston's Lying-In Hospital. The amniotic membranes were stripped from the placentas and then stirred into a solution containing trypsin, a digestive enzyme that detached the cells from the membrane. The cells were centrifuged and placed in a nutritious medium before being incubated. An essential ingredient in the nutritious medium was beef embryo extract, prepared from minced cow embryos obtained from a local slaughterhouse.

"Every day Enders took time to make the rounds of the laboratory benches to talk with each fellow and ask, 'What's new?'" Katz says. "That two-word question was a wonderfully effective stimulus to enhanced laboratory productivity because a fresh answer to the inquiry earned extra personal time with the Chief to discuss one's observations."

Isolating and Growing the Virus

Enders sent Tom Peebles, a pediatric resident, to draw blood and collect throat swabs from kids in Boston who had measles. Using roller tube techniques, the team began trying to isolate and grow virus from their samples. Eventually, they had nine cultures growing in human or monkey kidney cells from nine different people who had measles in different geographical areas and at different times. They continued to grow the viruses in serial passages through these media. From four to ten days after they were inoculated, the tissues began to show the abnormal, cytopathic changes that indicated a virus had taken over the cells. The cells would develop multiple nuclei and become enlarged, then die. Over time, nearly all of the cells followed this path to destruction.

They next combined human sera from twelve people who had been afflicted with measles to the tissues to see if antibodies in their sera would inhibit the cytopathic process. It did. This meant they were dealing with the measles virus and that the human immune system reacted successfully to the virus. As early as 1954, the tissue culture techniques were producing great results.

The Monkey Riddle

The next step for Enders and his colleagues was to inoculate monkeys with the virus they had grown after one, two, and twenty-three serial tissue passages. Just as previous researchers had found, the results varied: Some monkeys showed symptoms and others did not. To try to solve this riddle, they examined their next set of experimental monkeys before running any trials on them, and found that twenty-two of twenty-four monkeys from three different laboratories already had complement-fixing measles antibodies in their blood. Where did they come from? The researchers next examined thirty-one monkeys just caught in the wild from Malaya and the Philippines—none had measles antibodies. They didn't know how the laboratory monkeys had become contaminated with measles antibodies, but that did explain the variable results. It also allowed monkeys to be used as reliable test animals, once the contaminated ones were weeded out.

Growing the Virus in Chick Cells

At the same time, the lab was trying to grow the virus in tissues other than humans, such as pig, cattle, or chick cells. In order to make a vaccine, it was important to find a nonprimate tissue in which to grow the virus. Otherwise, there was a high risk of contamination by other agents that would be pathogenic to humans, such as viruses.

Month after month, these attempts proved unsuccessful, so time after time, Enders and his team left the lab without making progress. Enders had the habit of taking a taxi home from the lab and got to know many of the drivers. As one of his beliefs in life was treating everyone equally, he showed them all respect. Katz says, "In twelve years of close daily association, I can recall only one person for whom he could not find some redeeming feature, and that man was indeed a rogue." One evening, on the trip home, the taxi driver asked Enders, "How is the progress of your research?" Enders couldn't pretend it was progressing well. The driver encouraged him to keep at it, "Perhaps you will discover something wonderful, like Dr. Salk."

Indeed. Two years passed as they continued experiment after experiment, serial passage after serial passage, mincing more beef for more tissue cultures. There is no definite recipe for growing a virus in foreign tissue or for developing a vaccine. Virologist Joel Warren noted, with a literary allusion John Enders would have relished, that it is not unlike the recipe found in Macbeth: "Round about the cauldron go, in the poison'd entrail throw…cool it with a baboon's blood, then the charm is firm and good."

Finally, in the team's third year of working with measles, they observed an interesting phenomenon. After passaging what they called the Edmonston strain (which Peebles had collected from the throat of a boy named David Edmonston) twenty-three times through human kidney cells, then fourteen times through human amnion cells, they noticed that some cells did not undergo the typical cytopathic change. Instead of enlarging and blowing up when infected, some cells became star- or spindle-shaped. As the researchers continued to passage the virus, these cells came to be the predominant manifestation of the virus.

The virus had obviously changed. Might it now grow in a chick cell? After passaging the virus twenty-three times through human kidney cells and then twenty-eight times through human amnion cells, the researchers cultured it in chick embryo amnion cells, chosen in the hope they would be analogous to human amnion cells. Once inoculated, the cultures were incubated at thirty-five degrees C. for nine days. Success! They could continue to passage the virus many more times through those cells and obtain the same cytopathic reaction as they had seen in human amnion cells. They next tried growing measles in other chick embryonic tissues, and that also worked. At long last, they had found a growth medium that might yield a viable vaccine.

Attenuation

In order to attenuate the measles virus, Enders and his team needed to put it through enough passages to weaken it so it would no longer infect humans, but not destroy so many that it would no longer produce enough viral antigens to create an immune reaction. (The antigens are the protein coat "signature" of the virus, and this is what allows the immune system to recognize a particular virus and send antibodies after it.) In the end, the lab developed two potential recipes. The best was from the Edmonston strain. It was produced after an additional six passages through chick embryos and fourteen passages in chick cell cultures.

Tests on monkeys showed that the attenuated virus differed from the infectious virus in two necessary ways. The weakened virus could multiply and cause infection in chick-cell cultures, but neither multiplied nor caused symptoms in monkeys. Most important, monkeys produced measles antibodies fifteen to twenty-three days after inoculation, which indicated that their immune systems would fight off the wild measles virus.

Enders' team now had to establish that their virus would work as a vaccine in humans, so medical doctors began tests on small groups of children. Between December of 1958 and March of 1960, they ran six sets of tests on children. The vaccine gave the children a mild infection but nothing worse, and the tests did not make them contagious. More importantly, it gave children immunity from full-fledged measles. They published their results in eight papers in the *New England Journal of Medicine* in July 1960. At a conference in New York, the eminent virologist and bacteriologist Joseph Smadel stood up and said, "John, you've done it again."

The one negative effect of the vaccine was that 30 percent of the children in the trials came down with a fever as a result of the shots. Joseph Stokes, a pediatrician at the Children's Hospital in Philadelphia, developed the practical solution of giving shots of human gamma globulin, containing measles antibodies, in one arm, and the measles vaccine in the other, but this was not going to work on a worldwide scale. So, as he had done with polio, Enders made his attenuated strain available to others to refine his work.

Here the vaccine developer Maurice Hilleman stepped in and applied the resources of the pharmaceutical company Merck, Sharp, and Dohme (now the giant corporation, Merck). Hilleman discovered that Enders' strain had become contaminated with a chicken leukemia virus, and he found a breeder in California who had developed a line of chickens that were immune to the virus. Illustrating once again the importance of the serial passage technique, Merck's team passed Ender's measles virus twenty-four times through primary human kidney cells and twenty-eight times through human amniotic cells, as Enders had done, but then upped it to twelve times through embyronated hens' eggs and fifty-seven times through chick embryo cells. The end result of Hilleman's labor, after passing thirty-five quality control tests, was a commercial vaccine licensed for use in 1963.

Eradication

Enders later wrote that his work on the measles vaccine was the most personally satisfying and socially significant of his career. Over the following decades, the measles vaccine spread all over the world. By 2000, more than 80 percent of the world's children were receiving it. By 2006, the number of measles cases had fallen from 106 million in 1963 to 20 million, and the death rate from over 6 million to 345,000. Amy Pearce calculates that the polio vaccine has saved more than 1 million lives, and over 113 million lives have been saved by the measles vaccine. The vaccine's benefits have also extended to the fight against global warming, as far fewer children get measles' notorious fever (and its typical five degree increase in temperature), thereby reducing humanity's collective body heat by some 430 million degrees!

For a disease to be considered eradicable, it must meet three conditions: Humans must be its only host; it must be possible to diagnose the disease effectively; and a reliable intervention must exist. In 1988, the World Health Organization, UNICEF, and Rotary International initiated a worldwide polio eradication effort, trying to make it only the second disease, after smallpox, to perish from the Earth as a result of humanity's efforts. Today, there are fewer than 2,000 cases a year, all in four countries where polio is still endemic: Nigeria, India, Pakistan, and Afghanistan.

Measles meets all the criteria for eradication, as well. Developed countries have shown that measles caseloads can be reduced to near zero, with near-total immunization coverage and effective responses to isolated outbreaks. As of this writing, the leading international public health agencies have not taken the additional step of calling for an all-out eradication campaign against measles, which is still the leading cause of vaccine-preventable death around the world. This is a shame, because every single measles death is an unnecessary death that could be prevented.

John Enders' work has touched billions of lives around the world. Within two decades of the publication of Enders' tissue culture techniques, vaccines for measles, mumps, rubella, and chickenpox became common. What's more, the polio publicity campaign, which used fear, celebrities, and hype to galvanize public attention, played a hugely beneficial role in convincing the public to vaccinate their children. There were and still are religious, pseudoscience, and conspiracy groups that fight mandatory vaccination. But vaccination quickly became a rite of passage for most parents and children. Currently, the government recommends fifteen childhood vaccines, including those mentioned above, plus rotavirus (a cause of diarrhea), diphtheria, tetanus, pertussis (whooping cough), HiB (a cause of various bacterial diseases), pneumococcal (pneumonia), hepatitis A and B, meningococcal (meningitis), and, in some cases, influenza. When John Enders started his work on polio, only three of these vaccines existed. The beauty of vaccines is that children never know the diseases against which they are vaccinated. They may also not know John Enders, the father of modern vaccines, but his work has prevented much tears and suffering.

A Man of Letters and a Man of Science

When Enders retired, he could count many successful research careers that began in his lab, and many were those who noted his personal charm and unwavering curiosity in the workings of the world. Robbins said, "Once engaged with him, you maintained your relationship. I've really worked with him for the rest of my life." Katz says, "He was a renaissance scholar, interested in all knowledge, not just medical science. He was my father-figure as well as my mentor!" Enders received a host of honors from his peers and the public, including a Lasker Award, *Time* magazine's Man of the Year, and membership in the National Academy of Sciences and the Royal Society of Britain. Not bad for someone who didn't even know what he wanted to do until he was in his 30s, and whose significant work only began at age 50.

On a September evening at their water front home in Connecticut in 1985, Enders was reading T.S. Eliot aloud to his wife, Carolyn. He finished and went to bed, and then quietly died. He was

88. At his memorial service, his friend, the bishop F. C. Laurence, said, "John Enders never lost his sense of wonder—wonder at the great mystery that exists and surrounds all of God's creation. This awareness is what gave him his wide vision and open-mindedness, his continued interest in all things new, his ability to listen, his humility in the presence of this great mystery, and his never-ending search for the truth." His widow said that John briefly revealed his heart when he told her, concerning how creation ran, "There must be a mind behind it all."

Lives Saved: Over 1 Million

Discovery: Polio vaccine

Crucial Contributors

John Enders: Guided the effort to develop tissue cell culture techniques for viruses.

Thomas Weller: Came up with several breakthrough ideas in the development of tissue cell culture techniques for viruses.

Frederick Robbins: Worked with Enders on developing tissue cell culture techniques for viruses.

Jonas Salk: Developed a killed polio vaccine, using tissue cell culture techniques.

Albert Sabin: Developed a live, attenuated polio vaccine that was used in most of the world, using tissue cell culture techniques.

Lives Saved: Over 113 Million

Discovery: Measles vaccine

Crucial Contributors

John Enders: Led the team that developed the Edmonston measles vaccine.

Samuel Katz: Worked on transferring the measles virus from hens' eggs into chick embryo cell cultures and on vaccine tests on monkeys and children.

Thomas C. Peebles: Worked on isolating the measles virus.

Kevin McCarthy: Worked on inoculating monkeys with a passaged virus.

Milan Milovanovic: Worked on passaging the measles virus through human cells.

Anna Mitus: Worked on passaging the measles virus through human cells.

Ann Holloway: Worked with the team throughout.

Maurice Hilleman: Working at the pharmaceutical company, Merck, he led the effort to reformulate Ender's measles vaccine into a commercial vaccine that could be used all over the world.

John Enders in his office (1955). Verner Reed/Time & Life
Pictures Collection/Getty Images.

Mark Twain often visited Enders' family when he was growing up.

"I mouth the strange syllables of ten forgotten languages, letting my spirits fall, my youth pass."
—John Enders, in his twenties, before he became a scientist

"The results we have just described appeared to be of basic significance since they left no doubt that the three known antigenic types of poliomyelitis viruses could be grown without difficulty apparently in a wide variety of human cells," John Enders' team wrote, upon discovering how to grow the polio virus in test tubes. This was one of the key breakthroughs that led to the polio vaccine and many others.

Enders didn't receive his Ph.D. until he was 33 years old. He was 50 when he started the viral research that would revolutionize vaccines. At that time, only three childhood vaccines existed. Now there are fifteen.

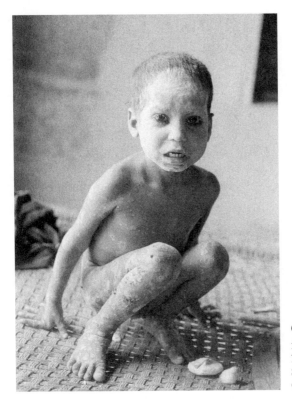

Child with measles in India.
Nik Wheeler/Sygma
Standard RM Collection/
CORBIS.

Measles is the most contagious common disease known.

Before a measles vaccine was available, more than nine-out-of-ten children caught measles in the United States, totaling more than 2 million a year. In 2004, there were 37 (all believed to be imported).

Eighty percent of the world's children now get vaccinated for measles, yet it is still the leading cause of vaccine-preventable death around the world, and the fifth leading cause of childhood mortality. When Enders began his work more than 6 million people died from measles each year. In 2006, the worldwide number was 345,000.

Humans are the only host for measles, so it could be completely eradicated.

Roller tubes. The test tubes containing viruses in tissue culture revolve slowly in the roller drums to simulate being in a live animal or human. Bettmann Standard RM Collection/ Bettman/CORBIS.

"Dr. Enders pitched a very long forward pass, and I happened to be in the right place to receive it."
—Jonas Salk

"The achievement of Enders, Weller, and Robbins was the starting point not only of modern poliovirology, but it launched the revolution rightly called molecular virology."
—Hans J. Eggers, MD

Producing a vaccine is exacting and arduous. Producing Enders' measles vaccine required the measles virus to be grown twenty-four times in human kidney cells, twenty-eight times in human amniotic cells, twelve times in embryonated hen's eggs, and fifty-seven times in chick embryo cells. After those 121 steps taken to weaken it, it still had to pass thirty-five quality control tests.

When Enders' team developed their tissue culture techniques only thirteen viruses that cause disease in humans had been identified. A decade later the number had risen to seventy-one, plus over 300 viruses that infect animals.

Chapter 7: Paul Müller— Over 21 Million Lives Saved

DDT and the Prevention of Malaria

THERE ARE over 3,000 species of mosquitoes, 130 of which live in North America, and they are wily creatures. Flying at one-mile-per hour, they sense exhaled carbon dioxide or infrared radiation to home in on warm-blooded animals like you and me. Only the female bites—she needs the protein found in blood to develop her eggs. She can deposit dengue fever, encephalitis, or yellow fever. But if you are unlucky or just in the wrong locale, there is one species, the Anopheles mosquito, which can inject you with a disease that dwarfs the others in lethality.

Perhaps you are drifting between consciousness and lovely sleep when you barely feel her proboscis poking through your skin. Unaware of her sucking your blood, you move your hand to brush the spot. She flies away, having consumed her weight in your blood, but she leaves behind sinister microscopic one-celled parasites. Within thirty minutes, the tiny parasites are infecting your liver. During the next day or two, all you notice is an itchy red bump needing an occasional nuisance scratch. But two weeks later, your liver cells burst open, releasing tens of thousands of daughter parasites. These parasites then infect red blood cells, releasing eight to twenty-four daughters with each cell burst. Unfortunately, your immune system has trouble attacking the invaders because the parasites attach to red blood cells with surface proteins of almost unlimited variation (each parasite can have as many as sixty different attaching proteins). The parasites cause red blood cells to bind to blood vessel linings and other red blood cells. As they accumulate, they jam together stickily, blocking blood flow in the body's smallest vessels and limiting the local oxygen supply. Now you have a fever, to be followed by sweats, headaches, and rigors. You need medical attention quickly, for you have malaria, one of the greatest scourges known to humankind.

Malaria's history mirrors that of human civilization. It is thought to have arisen in primates in Africa and evolved with humans, spreading with them as they migrated around the world. Evidence of its characteristic fevers appear in Chinese writings as far back as 2700 BC, and in the records of every major civilization since. Malaria kept the ancient Romans inside at night lest they catch what they called "the Roman fever"; in fact, the Romans are responsible for the disease's modern name, calling it "mal-aria," or bad air, from whence they thought it originated. Malaria has killed numerous popes throughout the centuries, prompting Pope Urban VIII to order in 1623 that a cure be found. He had good reason for his declaration—he'd been elected pope after the death of his predecessor from malaria, and his election occurred only after the majority of the cardinals themselves fell ill with the disease. Pope Urban spent the early weeks of his papacy shivering with malarial fevers, delaying his coronation by eight weeks. And malaria left enough of an impression on William Shakespeare that the playwright inserted references to it in several of his plays, including *The Tempest*: "All the infections that the sun sucks up / From bogs, fens, flats, on Prosper fall and make him / By inch-meal a disease!"

Because the disease has infected a large portion of the human population for thousands of years, forcing numerous genetic adaptations, Oxford University professor Dominic P. Kwiatkowski calls malaria "the strongest known force for evolutionary selection in the recent history of the human genome." People from Africa and other highly malarial areas have adapted by developing specific physiological characteristics that provide some protection against malaria. One is the sickle cell genetic trait. While the mutation doesn't protect against the initial infection, it does provide about 60 percent protection against mortality, most of the benefit being accrued to babies aged 2–16 months. Thus, as many as 40 percent of people in some areas of Africa and about 8 percent of all African-Americans carry the sickle cell trait. Unfortunately, if two people with this genetic recessive trait have a child, there is a 25 percent chance that the child will be born with sickle cell anemia, a painful, chronic, and sometimes fatal blood condition.

Another evolutionary response to malaria is the absence in some people of Duffy antigens, which are pairs of proteins that reside on the outside of red blood cells. Lacking them confers resistance to P. vivax, one of the four types of malaria, since that species uses the Duffy antigens to invade blood cells. European, Asian, and American populations have these antigens, but most West and Central African populations lack them, presumably because the adaptation conferred resistance in their hereditary lineage.

Until the early 1630s, these evolutionary responses were the only protection against malaria for Europeans. Then a Jesuit priest carried some Peruvian bark used by the Andean Indians to cure "shivering" back across the Atlantic Ocean to Rome. The "Jesuit powder," as it became known, was, of course, the earliest form of quinine. Although quinine works well, treatment must be obtained quickly. For strains of malaria that have become resistant to it, there are now other drugs available, such as artemisinin. However, only 60 percent of those suffering from malaria today have access to affordable and appropriate treatment within 24 hours of onset of symptoms. And no drug had an effect on the spread of the scourge itself, about which little was known until 1897. It wasn't bad air that brought on malaria; nor was malaria contagious. Army surgeon Ronald Ross discovered that the only way to catch malaria was through the mosquito. Ross's discovery of the vector allowed the first concerted approach against the disease—eliminate the mosquito, and you eliminate malaria. It was such an important discovery that in 1902 Ross became the first of four researchers to be awarded a Nobel Prize for work associated with malaria.

It may be easy for us today to dismiss malaria from our thoughts as an irrelevant tropical disease residing in the "Heart of Darkness"-type jungles of Africa. However, at least eight U.S. presidents had the disease, including both Washington and Lincoln. Prior to 1950, malaria was common in the southern United States, infecting 15,000 people a year and killing about the same number as scarlet fever. Malaria thrives wherever the Anopheles mosquito lives, so it was common in temperate climates all over the world, occurring even as far north as the Russian town of Archangel on the Arctic Circle. Today, it is endemic in parts of Mexico, Central and South America, the Middle East, and all of South and Southeast Asia. The dreaded parasites the mosquitoes carry are actually protozoa, which are in the kingdom *Protista*, a conglomeration category where scientists put organisms they don't know what to do with. They don't fit into the better-known classifications of animal, plant, fungus, and bacteria.

As a parasite, malaria lives on or in another living organism, to the detriment of its host. While it thrives in many animals, only four types infect humans. If you are infected with any of three types, you will likely survive, although the symptoms can reappear years later (the record is thirty years; the parasite hides out in the liver during this time and can reactivate). However, if you are infected with *Plasmodium falciparum*, the most common species in Africa, your sickness may be severe or even deadly. It is especially fatal to children and pregnant women. Each day in sub-Saharan Africa, some 2,000 children perish from the disease's onslaught.

In one study, it was found that the average person with malaria had had it twelve times before. This recurring sickness erodes people's productivity; it is estimated to cost the African continent $12 billion a year in lost gross domestic product because of its debilitating symptoms. Today, malaria still infects as many as 500 million people each year, killing more than a million. Forty to 50 percent of the world's population live in areas held hostage to the Anopheles mosquito's piercing sup.

Given these present-day statistics, it is hard to believe that things could have ever been worse, but at the turn of the twentieth century it was estimated that 10 percent of all deaths around the globe were caused by malaria. Then along came a man who completely changed that equation. He discovered a chemical so portentous that in a 2005 essay British politician Dick Taverne called it "the single most effective agent ever developed for saving human life." What he discovered has gone on a roller coaster ride of public opinion. At first sanctified, it would within a few decades be considered one of the most evil demons man had ever created, right up there with atomic weapons. Today, it hangs in the balance: Was it a colossal mistake, or can it be redeemed?

A Boy's Fascination with Chemistry

Paul Hermann Müller, the eldest of four children, was born in Olten, Switzerland, on January 12, 1899, outside Basel on the Rhine River where Germany, France, and Switzerland meet. His father was a railway clerk and his Lutheran mother ran the family with an iron hand. At first a mediocre student, Müller suddenly awoke to the pleasures of experimentation when he discovered science and chemistry in high school. Unfortunately, his fascination didn't help his grades. Instead of studying, he spent hours locked in his home laboratory working on his own experiments, with his father's encouragement. Despite constant lecturing from his mother and his school principal, the frustrated Müller dropped out of school at 17 during World War I. He found work at industrial chemical companies in Basel.

During this time, two events occurred that would influence the rest of his life. The first was a severe war-exacerbated food shortage in Switzerland. The second event was the Russian typhus epidemic, the greatest such epidemic in history. It resulted from the chaos of the Russian Revolution, coupled with a lack of any effective insecticide against the common louse, which carried the infectious organism. Up to 30 million people contracted the disease and nearly 3 million died.

After two years in the "real world," Müller had matured enough to return to high school. In 1920, with his diploma firmly in hand, he entered the University of Basel. The ambivalent high school student disappeared under the tutelage of chemistry professor Fichter. Müller became a hard-working student, spending all his free time working in the labs both as a student and a paid assistant. He had dark hair and rather large ears, and he was so thin he was known as "The Ghost." He majored in chemistry, with minors in physics and botany. Professor Fichter proved to be both a mentor and a friend, and he encouraged Müller to go on to graduate school. Müller had proven very adept at chemistry and Fichter, thinking practically, assigned him a Ph.D. thesis on a compound used to manufacture dyes, since that was what the nearby chemical companies produced.

In 1925, after completing his graduate work at Basel, Müller joined the chemical company J.R. Geigy (which eventually became today's drug giant Novartis). His initial work focused on producing synthetic tanning materials. He then developed a seed disinfectant that wasn't based on mercury, which, though poisonous, was at the time widely used in agricultural products. As the decade of the twenties ended, Müller married. He and his wife, Friedel, eventually had two sons, Heinrich (b. 1929) and Niklaus (b. 1933). After the birth of their last child, their daughter Margaretha in 1934, an opportunity presented itself. Geigy wanted someone to work on developing new insecticides. In those days, saving fabrics and raw materials from the damaging effects of insects was of major economic import and since the company had already invented a compound that protected woolens from moths, it now wanted to branch out into eliminating damage from other insects. Other organic chemists felt insecticides were beneath them, but Müller, whose interest in botany made the task compelling, volunteered. So in 1935 he began a deliberate search for new insecticides.

It was the perfect job for Müller. In her book *Prometheus in the Lab*, Sharon Bertsch McGrayne writes that "as a scientist Müller was doggedly determined and methodical, his perception was acute, and his outward reserve covered a passionate devotion to science." He began by reading background academic papers relevant to his research. His daughter, Margaretha, remembers that he often was so absorbed in his work that he seemed to be wearing blinders that blocked everything else from view.

Although consumed by his work in the lab, Müller did have a life outside the walls of Geigy. One of his greatest joys was spending time in the weekend cottage the family owned in the Jura Mountains. Lying in northwest Switzerland, the mountains are more similar to the Blue Ridge Mountains of North Carolina than the taller and better-known Alps. The area had been home to the Swiss watchmaking industry since 1700. There, Müller alternated between reading science articles on pest control and breaking away to indulge his hidden naturalist, taking his children on nature walks, gardening, and photographing the wildflowers near the cottage.

Part of Müller's reading focused on insecticides already in use. The most common ones used in Europe and the United States were based on the doubly toxic lead arsenate, whose residues could poison people long after application. Müller was determined to find a safer substance.

The "Perfect" Insecticide

Müller shaped his own requirements for a successful synthetic insecticide. As he recounted in his Nobel speech, "I considered what my ideal insecticide should look like, and the properties it should possess. I soon realized that a contact or 'touch' insecticide would possess very much better prospects than an oral poison." The properties of this ideal insecticide should be as follows:

1. Great insect toxicity.
2. Rapid onset of toxic action.
3. Little or no mammalian or plant toxicity.
4. No irritant effect and no or only a faint odor (in any case not an unpleasant one).
5. The range of action should be as wide as possible, and cover as many Arthropods as possible.
6. Long, persistent action, i.e. good chemical stability.
7. Low price (= economic application).

Of the five known insecticide classes at the time, none met his criteria. Nonetheless, German scientists had been pursuing a plethora of synthetic compounds for years, and had already filed patents for dozens (some of which would later be used to eliminate humans as well as insects during World War II), although none were being sold at the time. Thus, Müller knew that in order for a new pesticide to receive a patent and succeed in the marketplace it would have to not only be cheap and effective, but also have a unique property that none of the existing products had.

Müller set about devising and testing various compounds. By now he was known in the labs as a lone wolf, or, as his daughter said, an *Eigenbrotler*—someone "who makes his own bread." He was no longer skinny, his white lab coat was often slightly askew, and his eyes would pierce the large glass beakers, tubes, and vials of the lab in long gazes. But he was not looking for a kaleidoscope at the end of the glassware; he was thinking of chemicals and how they might undergo a meta-morphosis in combinations. He was at home in his lab, testing all the compounds himself, spraying them into a big glass chamber and put-ting insects—primarily *Calliphora vomitoria*, otherwise known as the common house fly—into the chamber to see if they lived. Müller had a hands-on experimental approach that he maintained throughout his life, believing that doing experiments himself provided him with better insights than he would get by assigning the work to others. "Such per-sonally conducted biological investigations stimulate the chemist in his work," he said, "and at the same time he learns, by his observations, to understand the problems and uncertainties of biological tests and is thus better able to appreciate the difficulties facing his colleagues who work in biology. Sometimes new and valuable discoveries may be made by small changes in methods of application, or again by the correct observation of apparently unimportant side-effects."

After a year of fruitless experiments on 100 compounds, Mueller realized that "new substances do not always fulfill expectations; on the contrary, they are often bad and only seldom better."

After another year of testing, Müller recalled, "After the fruitless testing of hundreds of various substances I must admit that it was not easy to discover a good contact insecticide. In the field of natural science only persistence and sustained hard work will produce results, and so I said to myself, 'Now, more than ever, must I continue with the search.' This capacity I owe probably to strict upbringing by my teacher, Professor Fichter, who taught us that in chemistry results can only be achieved by using the utmost patience."

Müller diligently pursued a new insecticide from 1935 to 1939, testing a new compound every four days or so. But he didn't randomly combine chemicals like some mad scientist. Instead, he allowed insights drawn from previous investigations of his and others to suggest new chemical combinations. He had begun with Geigy's moth repellant—a chlorinated hydrocarbon with one or more carbon-chlorine bonds—because of its chemical stability. Later, he learned that compounds containing the chloromethyl group (carbon bonded to one chlorine and two hydrogen atoms) "showed a certain activity"—another piece to the puzzle. Finding a paper published in 1934 on a substance called diphenyl-trichloroethane inspired Müller to investigate related substances, which have a trichloromethyl grouping (carbon bonded to three chlorines).

Finally in September of 1939, after four years of work and 349 failures, Müller placed his 350th compound—a diphenyl-trichloroethane derivative—in the fly cage. For a short while, the flies buzzed around normally, and Müller sighed at what appeared to be another failure. But, looking at the cage after some time had elapsed, he was amazed to see flies falling helplessly out of the air. Time after time Müller tested the new substance, and always the flies died. Even more impressive was the compound's lasting power. As Müller recalled in his Nobel speech, "My fly cage was so toxic after a short period that even after very thorough cleaning of the cage, untreated flies, on touching the walls, fell to the floor. I could carry on my trials only after dismantling the cage, having it thoroughly cleaned and after that leaving it for about one month in the open air." (Later experiments would show that laboratory surfaces sprayed with the substance and kept free of dust were still toxic after seven years.)

Soon Müller confirmed that his insecticidal compound didn't need to be ingested, but could kill insects that simply came into contact with it. Thus the tested compound met all of Müller's requirements that he had set out in the beginning except one – it did not kill quickly. As a result, other biologists at Geigy weren't overly impressed, assuming that the slow onset of action indicated a lack of toxicity. They were used to insecticides that killed instantly upon ingestion, and doubted that Müller's compound would work in the field. But Müller, the baker of his own bread, continued testing, finding that his compound killed every insect he stuck in his glass cube in experiment after experiment.

Endless Intricacies of Nature

Müller prepared his compound from chloral, one of the first sleeping medications. It is the drug behind the sedative "Mickey Finns" that characters in old movies slip into the drinks of people they want to incapacitate. He also used the solvent chlorobenzene, with sulfuric acid as a catalyst. The product was dichloro-diphenyl-trichloroethane, which would become so well known by its acronym, DDT. Müller went back to the literature and found that he wasn't the first to combine those three chemicals to create this new chemical. The feat had been accomplished by an Austrian graduate student back in the 1870s, but that young student, Othmar Zeidler, never realized the compound's devastating effect on insects and so his mixture was forgotten. The fact that Müller later made an identical discovery independently from a predecessor is not uncommon in science. The accumulation of knowledge over time generates a constant flow of new ideas to scientists, so it is not surprising that more than one scientist can have the same insight. However, Müller was not only looking for a new chemical, he also was looking for a chemical that solved a specific problem—unwanted insects.

Scientists later discovered that DDT kills insects by binding to sodium ion channels in the insects' nerve cells, or neurons. These ion channels, which can open or close depending on the voltage difference between the inside and outside of the cell, allow neurons to conduct electrical signals. Normally, a discrete electrical signal travels the length of the neuron and is then passed along to other neurons via chemical neurotransmitters, after which the neuron remains inactive for a certain amount of time until it is excited by another electrical signal. This type of controlled signal conduction is the basis of the nervous system, both in insects and in other types of animals, including humans. When DDT binds to an insect's sodium channel, however, it forces the channel to remain open, causing the neuron to continue firing and eventually leading to paralysis and death of the insect. Fortunately for humans and other mammals, the sodium channels in our brains evolved slightly differently from those in insect brains, making them less susceptible to DDT. In addition, DDT is more potent at the body temperature of a typical insect than at human or mammalian body temperature and, furthermore, our bodies metabolize more of it before it reaches our brains than do insect bodies. That said, it is worth noting that DDT is very soluble in body lipids, and has the potential to affect a number of other systems in the body, so the fact that it does not target our sodium channels doesn't entirely let it off the toxicological hook.

Müller knew nothing of ion channels while he was developing DDT, or afterwards; he just knew that some chemicals are capable of killing insects while leaving other animals apparently unharmed. In fact, scientists today use DDT and other insecticides that act on insect sodium channels to study the tinkering that evolution has done with ion channels since the time that the insect and pre-mammalian genetic lines split. Thus, DDT illustrates a very common phenomenon in science, in which a discovery is made long in advance of a full understanding of the system to which it applies, and, in fact, the discovery itself opens up the way to greater knowledge of the endless intricacies of nature.

Müller continued testing his new chemical on different arthropods. At the time, the devastating Colorado potato beetle was threatening Switzerland's potato crop, so Müller demonstrated that a single application effectively killed the beetle and its larvae for up to six weeks. With a potential commercial use, the higher-ups at Geigy proceeded to patent DDT as an insecticide, then ran numerous experiments on different substances they could combine it with to deliver it to crops, animals, and humans. Geigy ramped up production and, by 1942, DDT was in wide use throughout Switzerland, saving the country's potato crop from those pesky beetles. In addition, Geigy provided the Swiss army commander with a ton of Neocid powder, full of DDT, to control lice that infested war refugees who were entering the country.

The Superior Insecticide of World War II

The story of how DDT reached the United States varies depending on the source. One version has it passing through a military attaché around 1941 as a "favor" from the Swiss. Ever neutral, however, the Swiss supposedly also passed it on to the British and Germans. What is known for sure is that in August of 1942 Geigy sent 400 pounds of DDT to its U.S. offices. No one bothered to open the package with its German instructions for weeks. Then a chemist translated the accompanying explanation and passed a sample on to the Department of Agriculture. The Department of Agriculture, in turn, sent it down to its research station in Orlando, Florida. The Orlando station had been charged with developing new pesticides to protect the troops. In addition to artillery and bombs, the Allies were under tremendous pressure from a much smaller enemy—insects. In 1942, for instance, the First Marine Division was pulled from combat in the Pacific and sent to Australia to recuperate after 10,000 out of 17,000 men contracted malaria.

The Orlando station was part of an effort that would eventually test and classify more than 13,000 chemicals for insect toxicity. But countless other scientists' formulations were no match for Paul Müller's creation. One application of DDT killed lice for a month, which was four times longer than previously available chemicals. More importantly, even a few specks of DDT floating in the air and landing on a beaker were enough to kill the mosquito larvae within. In a larger test, researchers sprayed a duck pond with DDT on a windless day to compare the effects on mosquito larvae in that pond with those in an untreated pond several miles away. A week after the test, mosquito larvae in the untreated pond died, killed by the DDT residue ducks carried from the treated pond on their feathers and feet. To ensure the chemical was safe for humans, volunteers slathered themselves with DDT and sat in vaults for hours, inhaling the fumes with no ill effects. Quickly realizing the potential of this new pesticide, the United States labeled it top secret, subject to censorship. Clandestine production began in late 1942 in dozens of factories.

The Allies first used the chemical to fight typhus in Italy. In January 1944, Army officials propelled a tiny puff of DDT powder down the necks and into the clothes of 1.3 million Italian civilians. This "dusting" protected them from typhus for weeks, for the first time in history averting an anticipated typhus epidemic. And not only did the chemical kill lice already present, but clothing treated with DDT became immune to them. Any lice that happened onto DDT-treated fabric died, and as soon as their eggs hatched, the larvae died, too. This effect lasted through several washings, just as had the DDT on Müller's fly cage. In the Pacific, an estimated 85 percent of the soldiers occupying the Bataan peninsula in the Philippines were infected with the disease. As a result, Air Force personnel began spraying DDT from the air onto island beachheads before invasions. When word reached officials that 500 men a day were falling sick from mosquito-borne dengue fever in Saipan in the spring of 1944, they carried out a DDT air strike, saturating the region with 9,000 gallons of diluted pesticide, allowing the Marines to take the island in June of that year.

Winston Churchill later espoused DDT's miraculous benefits when he proclaimed: "The excellent DDT powder, which has been fully experimented with and found to yield astonishing results, will henceforth be used on a great scale by the British forces in Burma, and the American and Australian forces in the Pacific and India and in all theatres." Eventually, DDT came to be considered the "wonder insecticide of World War II." Thanks to its use, the war was the first in modern history in which fewer people died from disease than from the war itself.

The Mosquito Men

In June of 1944, censorship of information on DDT was lifted and *Time* magazine ran an article about the chemical. It quoted Lieutenant Colonel A. L. Ahnfeldt of the U.S. Surgeon General's office as stating that "DDT will be to preventive medicine what Lister's discovery of antiseptics was to surgery." The reference was directed at malaria, against which preventive measures were direly needed. After Ronald Ross's discovery that the mosquito was the vector for malaria, a worldwide effort began to drain swamps, eliminate standing water, and kill as many anopheles mosquitoes as possible. But as far as chemical protection before DDT was concerned, the "mosquito men," who called themselves malariologists, had to make due with chrysanthemum-derived pyrethrum, which only killed mosquitoes in the room when it was sprayed, and diesel oil mixed with the toxic arsenic-based Paris Green mixture for bodies of standing water. When DDT came on the market in the mid-1940s, mosquito men had for the first time a vanquishing tool.

The most famous mosquito man was Fred Soper, the so-called "General Patton" of entomology. After working to eradicate yellow fever in South America, Soper proposed that with DDT it was possible to completely eradicate malaria. His focus was not on killing every anopheles mosquito. Rather, he supported spraying the insides of houses in infected zones. After imbibing blood from a human, mosquitoes will often land on a nearby wall to rest. There the DDT would kill them. Spraying all over a particular country could create a hiatus during which little malarial transmission would occur. It takes three to four years for a human to completely clear the parasite from his or her body, but once this time had passed, even if mosquitoes were allowed to propagate again due to decreased spraying, there would be no protozoan for them to ingest and pass on.

This program began with some stunning successes. The first occurred in 1946 in Sardinia, then the most malarial region of Italy. Soper attacked it with the vengeance of the general after whom he was nicknamed. With financial support from the Rockefeller Foundation, he hired 33,000 people to spray more than 286 tons of DDT in 337,000 buildings. Along with environmental approaches, such as draining swamps, he managed to reduce the number of malaria cases on the island from 75,000 in 1946 to nine in 1951. Such was the power of DDT.

A similar effort was made in the U.S. South. During the war there existed the Malaria Control in War Areas Agency. After the war, it was turned into the Communicable Disease Center and oversaw the U.S. National Malaria Eradication Program. Beginning on July 1, 1947, over 4,650,000 houses were sprayed in thirteen southern states. Trucks crisscrossed towns in the Deep South on hot, muggy nights, spraying vast clouds of the DDT insecticide. Children danced in it, waving their arms to make the fog swirl. By 1951, malaria was completely eradicated from the United States. The Center had little left to do, so it took up other health problems, eventually becoming the Centers for Disease Control (CDC), one of the preeminent health organizations in the world. Its offices are located in Atlanta because of its original geographical purpose of ridding the South of malaria.

These types of results led to the Global Malaria Eradication Program. The fledgling World Health Organization led the effort, with the United States providing $1.2 billion, half the funding. It was a grandiose plan, complicated by economic, logistical, and cultural problems. The early successes were so amazing that scientists predicted that the academic literature of agricultural and medical entomology would have to be re-written. In India in 1953, when the eradication campaign began, there were 75 million malaria cases a year and 800,000 deaths. By 1966, there were fewer than a million annual cases of malaria and zero deaths. A 1970 article in an Indian newspaper attributed the lengthened lifespan of people in that country from 32 years in 1948 to 52 years in 1970 to DDT.

In neighboring Sri Lanka (then known as Ceylon), DDT cut the incidence of malaria from about 3 million cases in 1946 to 7,300 cases in 1956, and the death rate fell to zero. Dr. Arthur Brown led the effort in parts of Southeast Asia and had the task of convincing the Buddhists to support the effort. He had met with a Buddhist monk who had informed him they would discuss it but might oppose it since they didn't believe in killing other living things, mosquitoes included, except for food. Dr. Brown relates that he was apprehensive when he went back to hear their decision: "A week later I mounted those imposing steps once more. Our young interpreter indicated the seat beside his master. We talked as we sipped orange juice, and at last we came to the problem of malaria. 'I am pleased to be able to tell you that we Buddhists will not oppose your program.... You see, Dr Brown, when the mosquitoes take their siesta on the wall, they are killing themselves. That, for us, is not a crime.'"

One after another, country after country experienced a malaria hiatus. In parts of Indonesia, 25 percent of the population were infected by malaria before DDT was available. When DDT was introduced, the rate fell to 1 percent. In Venezuela, the number of malaria cases dropped from 8 million to 800. Removing the vector for a long enough period of time wiped malaria out of humans' collective blood streams.

220

DDT didn't only work on mosquitoes, though; it worked on hundreds of pests. Industry and public officials quickly came to the belief that if a little DDT was good for humankind, then more had to be better. Soon it was being sprayed on millions of acres of forest to stop spruce budworm and on cotton fields by crop duster planes to control various pests. Rice seeds were soaked in it, and apple orchards were sprayed; it worked on the gypsy moth; in fact, it seemed to work on any pest. An underreported story is DDT's effectiveness on sandflies. These tiny insects, no bigger than the period at the end of this sentence, carry the parasitic protozoan which causes black sickness, or *kala azur*, a usually fatal disease that results in anemia, fevers, and an enlarged spleen, liver, and lymph nodes. The disease is endemic in most tropical and subtropical regions worldwide. Sandflies tend to reside on the floor and walls of houses. But just one application of DDT kept a home free of sandflies for a year. In fact, as long as spraying for malaria was ongoing in India, black sickness disease disappeared.

Soper's success and the World Health Organization's far-reaching goals came at a time when people began to think science's accomplishments might be unlimited. Penicillin, the first antibiotic, had just been discovered, and blood transfusions were coming into broad use. It seemed that science would allow humankind to control nature. In 1948 Paul Müller received the phone call scientists dream of—he had won the Nobel Prize in Medicine, despite the fact that he was neither a doctor nor a medical researcher, but rather a chemist. Such recognition speaks volumes about the world's perception of the benefits of DDT in preventing human disease. Müller and his family traveled by train north through Germany, passing through many towns still in ruins from World War II, toward the award ceremonies in Sweden. At the ceremonies, Gustaf Hellström, a member of the Royal Academy of Sciences, addressed the laureate: "After having tested different chemical combinations, you found one which killed, not flies alone, but also many other kinds of vermin, and with that you have made one of the greatest discoveries within the recent history of prophylactic medicine. DDT... kills the mosquito, which spreads malaria; the louse, which spreads typhus; the flea, which spreads the plague; and the sandfly, which spreads tropical diseases."

"To few chemicals does man owe as great a debt as to DDT," the U.S. National Academy of Sciences concluded in 1970. Amy Pearce estimates that DDT saved more than 21 million lives while it was widely used and it surely prevented many times that number of illnesses. In two decades, Paul Müller's new insecticide had gained international fame as a stupendous panacea of preventative medicine, but this fame would soon turn to infamy as DDT became the primary target of a burgeoning environmental movement, led by one marine biologist turned anti-pesticide crusader.

The Ecology of Fear
and the Birth of Environmentalism

In 1962, Paul Müller retired to work in a private laboratory he established, trying to find a pesticide that plants could absorb through their roots, rather than their leaves. Ironically, in that same year, Rachel Carson's book *Silent Spring* sprang. After being serialized in *The New Yorker*, it was published as one of the first major environmental treatises. The book was an indictment of the environmental devastation wrought by DDT, tracking the way DDT entered the food chain, and claiming it caused cancer and other damage. Carson warned that DDT would silence forever the birdsong and other natural sounds of our world. It was politicized when President Kennedy called for an investigation of the book's claims, while at the same time the chemical industry, eager to sell more insecticides, promoted their use. In this climate, *Silent Spring* became one of the most influential books of the 20th century, and is still assigned today to middle and high school students around the country.

The book centered upon the one negative attribute of DDT—its ability to dissolve well in oil. This characteristic didn't seem all that important at first, but as Rachel Carson pointed out, it turns out to be DDT's Achilles heel. It allows the substance to build up in the fatty tissue of animals, becoming more concentrated as it moves up the food chain. This wasn't completely surprising to Müller. After only a few years' use of his invention, the chemist in him could see the dangers. In his 1948 Nobel Prize lecture, he warned people to beware of humankind's ignorance: "Long years of patient, detailed study have produced explanations of the constitution of important vitamins, hormones, and bacteriostatic substances such as penicillin….Yet in spite of all these results, we are still far removed from being able to predict with any degree of reliability the physiological activity to be expected from any given constitution.... More difficult still are the relationships in the field of pesticides, and in particular of synthetic insecticides.... We are moving into unknown territory where there are no points of reference to begin with . ... . We can proceed only by feeling our way."

Rachel Carson had felt her way to a firm conclusion. She said that indiscriminate aerial spraying of DDT and other chemicals over millions of acres of forests, farms, and even towns was leading to disaster. Some of the facts backed her up. Müller had seen it firsthand in 1945 when he was invited to the United States to see how his invention was being used. He was horrified to see that rather than recognizing, as he had, that a little would go a long way, the U.S. had crop dusters blanketing fields with the stuff. The chemical was being used in much greater amounts than needed, and much more often than recommended. Müller was even sprayed with it at the Swiss border when he returned home from his U.S. trip. As the years passed, usage kept increasing. More than 80 million pounds of DDT was sprayed over the US in 1959 alone, almost half a pound per person.

The mass aerial spraying on forests and crops had a devastating effect. Mosquitoes became resistant to its toxicity. If it had been used sparingly, and limited to the interior walls of homes, resistance would not have occurred so rapidly. Still, it seemed that DDT influenced even resistant mosquitoes' behavior due to its "excito-repellent" effect, which causes mosquitoes to actually fly away after just smelling the insecticide. And not all insects developed resistance; sandflies, for example, never developed the same resistance to DDT as mosquitoes.

Rachel Carson went further than Müller in urging caution, and tarred DDT with the negative attributes of all poisonous insecticides. When she mentions other chemicals in the book, she compares their toxicity to DDT, some of which are hundreds of times more toxic. DDT was an acronym that the public could remember, so it became synonymous for all dangerous pesticides. Then, to culminate her argument, Carson associated DDT with cancer. "The assertion that pesticides were dangerous human carcinogens was a stroke of public relations genius," wrote Ronald Bailey in Reason Magazine in 2002. "Even people who do not care much about wildlife care a lot about their own health and the health of their children." There were no good studies of DDT causing cancer, but it was an alarmist era during which insidious associations played well.

Rachel Carson died of breast cancer in 1964, and a year later Paul Müller died of a stroke. His death proved bittersweet for his family who, while mourning his loss, were glad he never had to endure the vitriol directed against his discovery in the years that followed. Over the next decade, fueled by Rachel Carson's book, the environmental movement grew. In 1960, environmental groups in the U.S. had barely 100,000 members. By the end of the decade, they boasted over a million. They joined in well-founded outrage over the previous decade's above-ground testing of atomic weapons that left clouds of radiation floating through the atmosphere, the air over the nation's cities being filled with smog, oil spills on America's coasts, the use of Agent Orange in Vietnam, and over Cleveland, Ohio's Cuyahoga River's oil-filled water catching fire. Another factor was the impact of the Apollo manned space mission to the moon. Pictures showing the tiny blue earth in the vast darkness of space emphasized the preciousness of life, encouraging people to become concerned for the stewardship of the planet. When the first Earth Day was celebrated on April 22, 1970, more than 20 million Americans participated. Political change quickly followed, with the passage of the Clean Air Act, Clean Water Act, Endangered Species Act, and the creation of the Environmental Protection Agency.

Then, in 1972, the U.S. Environment Protection Agency banned DDT.

A Polarizing Life-or-Death Debate

Under Richard Nixon's administration, the EPA banned DDT despite seven months of hearings and the declaration of its own administration law judge that DDT did not cause cancer or birth defects in humans and did not have any significant deleterious effect on "freshwater fish, estuarine organisms, wild birds or other wildlife." The EPA's decision marked the first time the "precautionary principle," a principle that places the burden of proof of the safety of an action on those advocating for the action rather than against it, was used to justify a comprehensive ban of a chemical. The same year the U.S. banned DDT, Carson's book was re-released with a new cover. Her publisher wrote: "No single book did more to awaken and alarm the world than Rachel Carson's *Silent Spring*. It makes no difference that some of the fears she expressed ten years ago have proved groundless, or that here and there she may have been wrong in detail. Her case still stands, sometimes with different facts to support it."

The U.S. ban brought swift action: U.S. industry, DDT's primary producer, ended production. Soon more than thirty other countries banned the chemical, as well. In 1976, the World Health Organization admitted its failure to eradicate malaria worldwide and greatly scaled back its program. Economics, mosquito resistance to DDT, organizational inflexibility, inadequate research, and the demonizing of DDT were all blamed. Even though it was an awesome success in temperate countries, helping to completely eliminate malaria in Europe, the U.S., Japan and Australia, DDT was still labeled a failure by environmentalists.

There was a small cry by a few that in demonizing DDT, the environmentalists had thrown out the baby with the bath water. But the many environmental groups railing against DDT bartered no middle ground, and drowned out the opposing voices. Nor were the environmentalists content to only ban it in their own countries—they wanted a worldwide ban. And with malaria wiped out of most of the wealthy nations, it was easy for them to gain political support, ignoring the silent tears of people who were once again losing loved ones to malaria. Environmentalists even successfully encouraged countries such as Norway and the U.S. to withhold some forms of aid to countries still using DDT.

The zealots won out for two decades, while the continent of Africa was still churning out malaria victims. In the 1990s, a third of all hospital admissions in Africa were for malaria (compared to virtually none in the U.S.). In the 1990s, the United Nations began trying to get the Stockholm Convention on Persistent Organic Pollutants Treaty (POPs) passed. It laid out controls over the production, import, export, disposal and use of twelve persistent organic pollutants, including DDT. Greenpeace, The World Wide Fund For Nature (WWF), and over 300 other environmental groups pressed hard for a complete ban of DDT. In 1999, Clifton Curtis, Director of the WWF's Global Toxics Initiative, said, "These chemicals are completely unmanageable and as long as they are used somewhere on Earth, nobody is safe."

For the next decade, all manner of incendiary rhetoric was traded back and forth, both sides absolutely sure in their self-righteousness. The environmentalists were certain they were saving the world, and those in favor of using DDT were certain they were saving humanity. The latter now had decades of proof of what the ban had wrought. In Sri Lanka, the country's malaria burden had shrunk from 2.8 million cases in the 1940s to just seventeen in 1965. Five years after the country stopped using DDT, the number of cases had risen to 500,000. In the 1980s, Madagascar stopped using DDT and immediately had an epidemic of malaria, with more than 100,000 dying. The heartfelt rage over the ban was summed up in Lisa Makson's *Front Page Magazine* article entitled "Rachel Carson's Ecological Genocide." Similarly, Michael Crichton, author of *The Andromeda Strain* and *Jurassic Park*, had one of his characters in the novel *State of Fear* say that banning DDT was "arguably the greatest tragedy of the twentieth century" and that the ban "killed more than Hitler." In light of these vitriolic attacks, some of which were directed against Rachel Carson herself, it is interesting to note what Carson actually wrote in *Silent Spring*: "It is not my intention that chemical insecticides must never be used." We will never know what her position on a complete DDT ban would have been, but some environmentalists inspired by her certainly pressed for one.

The arguments led to research. One common environmental assertion was that there were more effective insecticides to control mosquitoes, usually citing deltamethrin. But in a U.S. Uniformed Services University study in Belize, huts treated with deltamethrin allowed in thirty-two times as many mosquitoes as those treated with DDT, and of those entering the DDT huts, most exited without biting. In fact, even forty years after the publication of *Silent Spring*, many of Carson's DDT claims have still failed to materialize. Although it was found to cause eggshell thinning in some bird species, DDT had no effect on others. While no one disputes that DDT and its metabolite DDE persists in the environment and in the fat of mammals and fish, there is no evidence of any significant toxicity in humans, nor any strong link with cancer. Dr. Donald Roberts, professor of tropical public health at the Uniformed Services University said, "You could eat a spoonful of it and it wouldn't hurt you." Amir Attaran, a lawyer and human rights advocate, went further, noting in an essay published in the *British Medical Journal* in 2000 that "not even one peer-reviewed, independently replicated study linking exposure to DDT with any adverse health outcomes exists." As negotiations on the treaty proceeded, one of the most effective campaigns came in 1999 from the Malarial Foundation International, an advocacy group started by Dr. Mary R. Galinski. It produced a public letter signed by more than 400 doctors from sixty-three countries advocating the continued use of DDT.

Finally, in 2001, DDT supporters achieved a breakthrough. The parties to the POPS treaty agreed to grant DDT a "health-related exemption" until cost-effective, environmentally friendly alternatives could be found. The positive effects in the fight against malaria were quickly apparent. In 2006, the *Malarial Journal* reported in a commentary: "Recently African countries that have reverted to DDT use have seen spectacular successes in their malaria control efforts. These include South Africa, Mozambique, Zambia, Madagascar, and Swaziland, who within two years of starting DDT programmes, slashed their malaria rates by 75 percent or more. With fewer people getting sick, access to ACT (an anti-malarial agent) drugs should be more feasible to nearly all victims, which should also cut malaria rates even further. Other African countries should learn from these shining examples, and start using DDT instead of sitting on the fence appeasing environmentalists who appear to care less about the lives of others."

In September 2006, nearly thirty years after phasing out the widespread use of DDT to fight malaria, the World Health Organization reversed its stance. As the director of the WHO's malaria department said: "Of the dozen pesticides WHO has approved as safe for house spraying, the most effective is DDT." Indoor spraying involves small amounts sprayed once or twice a year at a cost of about $5 (most of that being for labor). No more is needed because the chemical has such a long staying power. Even though some mosquitoes have developed a resistance to it, they still are repelled from DDT-sprayed environments. And when limited to indoor use, it affects few birds or other animals.

Finally, even the environmentalists came around to the evidence. "South Africa was right to use DDT," said WWF spokesperson Richard Liroff in 2005. "If the alternatives to DDT aren't working, as they weren't in South Africa, geez, you've got to use it. In South Africa, it prevented tens of thousands of malaria cases and saved lots of lives." Greenpeace spokesperson Rick Hind agreed: "If there's nothing else and it's going to save lives, we're all for it. Nobody's dogmatic about it."

Steve Milloy, an adjunct scholar at the Cato Institute, thought the environmentalists' change of mind was belated: "It might be easy for some to dismiss the past 43 years of eco-hysteria over DDT with a simple 'never mind,' except for the blood of millions of people dripping from the hands of the WWF, Greenpeace, Rachel Carson, the Environmental Defense Fund, and other junk science-fueled opponents of DDT."

Humankind's struggle against malaria, public hysteria, and dogmatism are far from over, but the one clear champion in this story is the scientific method. At each step along the way, evidence accumulated by scientists has enhanced and refined our understanding of DDT and its context in the environment and the society in which we live. Science itself does not insist on seeing issues in black or white, and it can be used to perform a nuanced risk-benefit analysis. Very few substances, even medicines, have zero risk when used by humans on a large scale. Science is based on evidence alone, not dogma, and when new evidence is discovered that casts doubt on old theories, science will adapt.

But it is up to humans and their institutions to interpret and act upon the evidence that science provides us, and as often seems to be the case, public and political opinion lagged significantly behind the scientific evidence both in the effort to scale back DDT's initial use and in the wrangling over an all-out ban. Rachel Carson correctly pointed out the evidence that indicated DDT used indiscriminately led to resistance and did harm the environment. But the evidence never incriminated DDT itself, just massive aerial spraying and use in agriculture. When it was finally accepted that DDT need not be demonized, that it could be used safely and beneficially in small amounts, indoors, away from birds (the amount of DDT put on 1,000 acres of cotton during a single growing season, for instance, can protect a whole country like Guyana from malaria for a year when applied indoors), the evidence overwhelmingly showed that DDT is the best insecticide currently available to prevent malaria by killing or repelling the anopheles mosquito. So today, DDT is back at work in some parts of the world, saving lives, just as it did during the 1940s and 1950s. Paul Müller can rest in peace, because his life's work is once again preventing tears.

Lives Saved: Over 21 Million

Discovery: DDT

Crucial Contributors

Paul Müller: Beginning from scratch, after four years of try-
                    ing to develop the ideal pesticide, discovered
                    DDT as his 350th compound. DDT proved to be
                    the safest, most effective insecticide of the era.
                    It killed the Anephelos mosquito, the vector that
                    infects people with malaria. It also killed other
                    disease-spreading insects.

Paul Müller
in 1948.
Bettmann Standard
RM Collection/
Bettman/CORBIS.

"My fly cage was so toxic after a short period that even after very thorough cleaning of the cage, untreated flies, on touching the walls, fell to the floor. I could carry on my trials only after dismantling the cage, having it thoroughly cleaned and after that leaving it for about one month in the open air."
—Paul Müller, just after discovering DDT

"The excellent DDT powder, which has been fully experimented with and found to yield astonishing results, will henceforth be used on a great scale by the British forces in Burma, and the American and Australian forces in the Pacific and India and in all theatres."
—Winston Churchill

"After having tested different chemical combinations, you...made one of the greatest discoveries within the recent history of prophylactic medicine. DDT... kills the mosquito, which spreads malaria; the louse, which spreads typhus; the flea, which spreads the plague; and the sandfly, which spreads tropical diseases."
—Gustaf Hellström, at the Nobel Prize ceremonies

Always independent, Paul Müller was known in the labs as a lone wolf, or as his daughter related, an *Eigenbrotler*, someone "who makes his own bread."

A 1945 spraying of DDT at Jones Beach, New York.
Bettmann Standard RM Collection/Bettman/CORBIS.

Prior to 1950, malaria was common in the southern United States, infecting 15,000 people a year and killing about the same number as scarlet fever.

Beginning in 1947, 4.6 million houses were sprayed in the United States, completely eradicating malaria from the country. Similar sprayings eradicated malaria from Europe.

The Center for Disease Control (CDC) began as an organization to eradicate malaria. That's why it's located in Atlanta, GA, in the southern United States.

In India, when the eradication campaign began in 1953, there were 75 million malaria cases a year and 800,000 deaths. By 1966 there were fewer than a million annual cases of malaria and no deaths.

In parts of Indonesia, 25 percent of the population were infected by malaria. When DDT was introduced, the rate fell to 1 percent.

In Venezuela the number of malaria cases dropped from 8 million to 800.

"DDT is the single most effective agent ever developed for saving human life."
—British politician Dick Taverne

DDT, thought by some to be a panacea, came to be sprayed indiscriminately. More than 80 million pounds was sprayed over the United States in 1959 alone, almost a half pound per person.
Loomis Dean/Time & Life Pictures/Getty Images.

Rachel Carson's book *Silent Spring* criticized DDT's effect on the environment and successfully galvanized an effort to ban it. It has been banned in much of the world since 1972.

"You could eat a spoonful of it and it wouldn't hurt you."
—Dr. Donald Roberts, professor, Uniformed Services University

"Not even one peer-reviewed, independently replicated study linking exposure to DDT with any adverse health outcomes exists."
—Amir Attaran, 2000 *British Medical Journal* essay

"If there's nothing else and it's going to save lives, we're all for it. Nobody's dogmatic about it."
—Greenpeace spokesperson Rick Hind, after Greenpeace stopped their effort to completely ban DDT

Today malaria still kills 2,000 children a day, most in sub-Saharan Africa. Recently, DDT has begun to be used again in parts of Africa. Walls are sprayed with amounts small enough to limit environmental damage.

Chapter 8: Howard Florey— Over 80 Million Lives Saved

Penicillin: The Miracle of Antibiotics

SEVENTY YEARS ago, an immense number of diseases were untreatable. Doctors fumbled through their black bags, groping among medicines based on mercury or arsenic that they knew were too toxic to be used in quantities that would work. All they really had to offer were vain words of comfort. Children were used to losing playmates to diphtheria, meningitis, rheumatic fever, or scarlet fever. In 1924, President Calvin Coolidge's youngest son got a blister playing tennis. The 16-year-old's blister became infected, the infection spread, and the resulting blood poisoning killed him. Even in the U.S., many women giving birth developed childbed fever. Adults commonly died of syphilis, sepsis, or pneumonia. These dangers disappeared quite suddenly in 1942 with the advent of a new drug that would revolutionize medicine: penicillin, the first immensely strong antibiotic.

No one even knew antibiotics were needed until the middle of the nineteenth century, when Louis Pasteur proved that invisible organisms all around us are the cause of many diseases. In the decades that followed, scientists realized that most of these "germs" fell into two broad categories: bacteria and viruses. Viruses are so tiny that they remained invisible until the advent of the electron microscope in the 1930s, but bacteria could be observed under the optical microscope, which had been invented centuries before, but was only then being perfected. Scientists have been studying them ever since. Bacteria are single-celled organisms that exist nearly everywhere—forty miles up in the stratosphere and miles below the ocean floor. Scientists at the University of Georgia have estimated that the number of bacteria on the planet approaches five million trillion trillion (that's a five with thirty zeroes after it). Each square centimeter of your skin carries about 100,000 bacteria, and most of us carry around over 500 bacterial species that add about three pounds to our body weight. In fact, there are about ten times as many bacterial cells as there are human cells in our bodies. Many are indispensable to us, ruminating in our gut to aid digestion.

While most bacteria are harmless, a few release toxins that attack our bodies. Some of these rogue bacteria, such as those that cause bubonic plague and cholera, are so infectious that they have been responsible for some of history's greatest epidemics.

Without a cure for bacterial infections, hospitals were very different prior to 1942 than they are today. Often people refused admission into a hospital as they were considered places to die. As the English doctor, Charles Fletcher, said, "Every hospital had a septic ward, filled with patients with chronic discharging abscesses, sinuses, septic joints, and sometimes meningitis…chambers of horrors, seems the best way to describe those old septic wards." Carbuncles oozing pus and abscesses on the body the size of a cup were not uncommon sights. And there was little in the way of treatment; about half the people who entered the wards exited on gurneys to the morgue.

A Great Fire Seemed to Burn within Him

It was into this era that Howard Florey was born. Florey's father, a boot-maker, emigrated from England to Australia in 1882 when his wife came down with tuberculosis. Sunshine and fresh air were thought to be the best treatment, so he moved his family to Adelaide, on Australia's south-central coast, where he established an initially prosperous boot factory. Sadly, his wife died four years later. Howard was born on September 24, 1898, nine years after his father remarried. The youngest of five children, his four sisters adored him, at times dressing him up in prissy, laced clothing so that with his long and curly hair he looked like Gainsborough's *Blue Boy*, a popular painting of the time. He grew to be handsome, if short (5'7"), an excellent athlete, competitive to a fault, and spiced with a fiery temper. Attending an upper-class private school, he acquired the nickname "Floss." In class he won prizes in science competitions and was most intrigued by chemistry, but the school's headmaster insisted there was no future for a chemist in colonial Australia. Florey instead went off to Adelaide University Medical School, which had only two dozen students. The school ground out practitioners, not researchers, and students were told to read the text and not ask questions.

While working on the Adelaide Medical Students' *Society Review*, Florey asked Ethel Reed, one of the few women in the medical school, to write an article on women in medicine, as an excuse to ask her to tea. Thus began what would turn out to be a long and difficult relationship. Later that year, Ethel came down with pleurisy, an inflammation of the lungs. Howard prescribed rest, "As I do for all my patients; it's the only treatment I'm sure about."

Upon graduation, Florey found the practice of medicine dramatically different from his studies. Since doctors had few cures, they concealed their ignorance by using big words and pompous phrases. He wrote, "I'm now 'Doctor' to the patients and I have to cover my ignorance by waving my arms and looking grave." In a short time he experienced the worst part of being a doctor, "the appalling thing of seeing young people maimed or wiped out while one can do nothing."

He dreamed of alleviating that ignorance by performing research, but to do so he would have to leave Australia. Thanks to his school prizes and athleticism, he received a Rhodes Scholarship to Oxford, among the most renowned universities in the world. To get to England, he found work as a ship's surgeon, which proved to be a great month-long adventure. Before he boarded, he wrote to Ethel, "The [ship's] tools for carving consist of some scalpels which might cut tobacco, some needles and a pair of artery forceps. I ordered a few extra but feel inadequate for anything big, e.g. disarticulation of head from neck." He tried for days to photograph the albatross that followed the ship, and he tasted his first rum.

In England, Florey entered Magdalen College, one of the most prestigious and class-conscious of Oxford's campuses. A colonial rube in the temple of upper-class England, he felt socially out of place. "The Englishmen are a queer lot," Florey reported to Ethel. "The chap on the same landing as myself went to Winchester…and has spoken [to me], but the majority preserve a frigid silence. I am assured that it is their manner and when you thaw it they are very decent. I'm also told it takes patience. They all seem to take themselves so seriously that I want to laugh."

While his fellow students treated him with the traditional disdain, he set himself to impress the faculty, a far more useful enterprise. One of those professors was Charles Sherrington, professor of physiology, who was an enthusiastic and inspiring teacher. Sherrington had Florey study the blood flow into cats' brains. "I've had a cat with a glass window in its head running around the lab, which amused me greatly," he wrote Ethel. He won a major fellowship to Cambridge, marking him as a rising star in British science, and gained Sherrington as a mentor. Florey was on cloud nine, writing, "Sherry has seen fit to let me work under him.... This is God's own opportunity. He's the President of the Royal Society and probably is recognized as the greatest living physiologist."

In 1925, Florey won a Rockefeller grant to study in the U.S., where he again impressed older scientists with his brilliance. He took as a mentor Dr. Alfred Richards of the University of Pennsylvania, who would become president of the National Academy of Sciences, and later a valued ally.

Back in Australia, Ethel had begun losing her hearing. She vacillated back and forth on whether to follow Howard to England, but after getting her medical degree, she finally agreed. When Florey returned to England he was reunited with her on his 28th birthday, and they married a month later.

In 1931, the chair of pathology at Sheffield University, 100 miles northwest of Cambridge, became open and Florey applied. It would take more than an excellent academic record to head a laboratory, but Florey was well suited to the task. As Gwyn Macfarlane, who knew Florey personally and would later write a biography of him, put it, "Florey's most striking characteristics were his energy and enthusiasm for research and his complete scientific honesty. He was a prodigious worker, full of ideas for the practical solution to some immediate problem, impatient of delay, and with an infectious vitality that was to attract a succession of collaborators who often found themselves doing the best work of their lives under his influence."

His electors instructed him, "We don't care what you do as long as you make a mess in this laboratory; it's been too clean for a number of years." Florey, who liked to perform one experiment every day, including Sundays, enthusiastically complied. D. E. Harding, a senior doctor at the university, said of him, "As a researcher he had more ideas, and better ones, than anyone else I have met. Almost all the work we did together, and some that I published alone, stemmed from ideas from Florey. With this went a capacity for hard work—one project involved us spending the nights as well as the days in the University. We did alternate nights for several weeks." Florey was usually running five or six lines of experiments, each with different collaborators, and he published academic papers widely, building his reputation.

This was the era when research funding was low, so Florey had plenty of administrative problems. For some he found practical solutions: He optimized the laboratory for research by dividing it into "cow stalls," much like the cubicles that are so common today. To get money, he became pushy. "One can get away with any sort of gaucherie as just one of these rough colonials," he said. "It's a very good line to play."

Through these years, Florey and Ethel had two children, Paquita, born in 1929, and Charles, in 1934; however, their marriage was otherwise rocky. They began living in separate rooms and, bizarrely, communicated by leaving notes on a table. In one note Ethel recommended divorce. Florey replied, "From the children's point of view, this would be lamentable. The children are now a most important link, and I want you to try to live with me, and let us both strive to rectify our mistakes as well as we may. Although neither of us have got all we expected from our marriage, there is still a good deal left which, with a little good will, we can make more. Let us try...."

In 1935, the chair of pathology at Oxford University came open. Getting appointed to a prestigious endowed chair at Oxford was a matter of high politics. As Florey biographer Eric Lax said, it was much like electing a pope. Luckily, Florey was still being supported behind the scenes by his early professor, Sherrington, who at 76 wanted to retire, leaving a good man in the post. Floery also had the backing of Edward Mellanby, the Secretary of the Medical Research Council in Britain, and one of the most powerful men in British medicine. Mellanby had seen Florey in action and comprehended the important work he was doing.

Six of the seven electors were present when they voted on the chair, and Florey lost the vote. Then, just before the meeting adjourned, Mellanby arrived, having been held up for two hours by a late train. Not timid in his opinions and quite persuasive—especially since he headed a council that distributed research money—he kept promoting Florey until a new vote was taken. This time Florey won by one vote.

The Need for Chemistry

Florey realized that at Oxford he would need to let go of his rough Australian temper in order to play the politics of academia, so he opted not to shake things up immediately. Instead, he began gradually transforming the school into his vision of what a research institute should be. He reassigned many of the present faculty to positions out of his way, and completely revamped the teaching courses, which had become a yearly repetition of lectures and unimaginative laboratory work that reminded Florey of his own education in Australia.

By now Florey had a strong belief that the study of disease would "not go very well without a very big injection of chemistry into it," and that in order to truly understand medicine, scientists needed to find the specific active chemicals within the compound substances that nature provided. He said, "I have never forgotten the remark made by Szent Györgyi, a Nobel Prize winner, in 1929, when I had the good fortune to work with him in Cambridge. He said that biochemical methods were then sufficiently good to enable any naturally occurring substance to be extracted, provided there was a quick test for it." Florey insisted that a chemist be on his research team.

A biochemist at Cambridge recommended that Florey hire Ernst Chain. Chain was the grandson of a Talmudic scholar in what is now Belarus, and the son of a chemist who had a successful company in Germany. "I was indoctrinated by both my parents," Chain wrote, "with a maxim that was beyond discussion, that the only worthwhile occupation in life was the pursuit of intellectual activities and any career which was not a university career was unthinkable." In 1933, disgusted by the rise of Hitler, he moved to England, leaving his family behind. At the time he thought Hitler's regime was a temporary aberration, but his mother and sister would later be murdered in the Holocaust. Chain spoke English with a pronounced German accent, looked like a well-groomed Albert Einstein, was voluble and flamboyant, and threw things when he lost his temper. In other words, he was exactly what the gentlemen scientists of England were not. Despite, or perhaps because of this, Florey and Chain got along famously in the first years, often walking home together in the evening.

When Florey began a study of the metabolism of tumors, he and Chain realized they needed someone who was proficient in micro-dissection. Chain recommended Norman Heatley, who had just written his doctoral thesis on the new field of measuring minute amounts of individual elements in biological substances. Florey hired him to be Chain's assistant, but within months Heatley and Chain, who was steeped in the German academic tradition that assistants be completely subservient to those higher ranked, began having trouble getting along. Heatley, a bright Ph.D. himself, felt that he "was perhaps better at actually doing research" than his boss Chain. Shouting arguments soon punctured the calm at the usually quiet pathology school.

Florey's biggest problem, however, remained the tight budget constraints on his research, caused by the slowly-fading Depression. He cut expenses to the bone, getting rid of the barrels of beer at meetings and even going so far as to ban use of the lift to save £25 a year. Over the next five years, Florey would hit up almost all of the organizations in England that granted research funds until he was sick of it. He wrote Mellanby: "The financial difficulties of trying to keep work going here are more than I am prepared to go on shouldering, as it seems to me that I have acquired a reputation of being some sort of academic highway robber because I have to make such frequent applications for grants from all sorts of places.... You may gather that I am fed up."

Nevertheless, the team continued their work. In 1937, as they were finishing the tumor research, Florey asked Chain and Heatley to start researching lysozyme. Alexander Fleming, a Scottish biologist, had discovered lysozyme in people's saliva and described it as an enzyme that could kill some bacteria. Florey had been researching it off and on since 1929. Interest in antibiosis, as the property of being able to kill bacteria was called at the time, had risen when the Bayer Corporation's Gerhard Domagk reasoned that because some dyes attach permanently to proteins in fibers of cloth, they might also attach to proteins in bacteria. Since dyes could be toxic, they might then kill the bacteria. In 1935, Domagk announced his group had made such a drug, which they named Prontosil. Researchers at the Pasteur Institute in Paris found that Prontosil's active ingredient was sulfanilamide. Because they were the first known drugs that had antibiotic characteristics, scientists became very excited about what became known as the sulfa drugs. Unfortunately, they only treated a few diseases and could have harsh side effects.

Chain proved that lysozyme was indeed an enzyme, a protein that accelerates and controls a chemical reaction, and he determined its chemical structure; but Florey realized that they would not progress beyond Fleming's earlier conclusion that lysozyme's antibacterial properties were useless against bacteria that seriously harmed people. Beatrice Pullinger, a doctor who worked with Florey in the mid-thirties, remembers afternoon teas with Florey and the medical doctors. They often told of hopeless bacterial infections that led to death, which would make Florey cringe. She believed these accounts influenced Florey's research direction. By 1938, as David Masters put it in a book based on interviews with Florey, "Florey had concluded that the anti-bacterial substances offered almost a virgin field for pure research."

Florey and Chain discussed performing a survey of such substances, and Florey recommended that Chain read a 1928 article entitled "Les Associations Microbiennes," which had a long historical perspective chapter on antibiosis that included a huge number of references. Chain was a voracious reader. In this era before search engines, reading scientific articles meant spending days on end in libraries, pulling out of the stacks musty journal articles. Beginning with the article's bibliography and then supplementing it with further reading, Chain plowed through about 200 references on substances known to inhibit the growth of bacteria. One day, Chain came upon a short, modest article Alexander Fleming had written in the *British Journal of Experimental Pathology*, dated 1929: "On the Antibacterial Action of Cultures of a Penicillium with Special Reference to their Use in the Isolation of B. Influenzae."

Fleming's Serendipitous Discovery

Alexander Fleming was 47 years old the summer of 1928, and had just been named Professor of Bacteriology at St. Mary's Medical School in London. He spent that August at his country home in Suffolk and had returned to his laboratory for a quick visit because one of his colleagues was treating a patient with a bacterial abscess. Fleming's lab was a bit of a mess. He had been culturing *Staphylococcus aureus*, a type of bacteria that causes numerous illnesses, including food poisoning, and the many dirty dishes he had left in a basin contained staph colonies. He was rummaging through the basin that August day when he noticed something odd. A plate, its bottom filled with staph, had been contaminated by a mold or fungus. What struck Fleming was a halo-like clear area around a growth of yellow-green mold. Something had killed off the bacteria.

Fleming put the dish down and went back on vacation. He wasn't the first to notice mold's antibiotic characteristics, but observation alone had produced little. When he returned, Fleming studied the mold for a while. Another laboratory worker identified it as a penicillium mold that likely came from a family of common ground molds being studied in a lab downstairs. A spore of the penicillium mold had escaped and floated up to Fleming's lab. Fleming showed that it was nontoxic in mice and, using human blood in test tubes, that it had a very short life—about a half hour—meaning that it was highly unstable. He found it hard to produce in sufficiently pure quantities to be useful even for testing, but he thought it might have some antiseptic uses if applied to open wounds, and that it might be used in the laboratory to weed out unwanted bacteria in mixed cultures. Whether Fleming thought that separating out and purifying the active substance in the mold would call for someone with a greater knowledge of chemistry than he had, or whether he thought it was simply too unstable to be useful even if isolated, he published his findings and then dropped the research. Penicillium remained nothing but a mold for the next ten years. It would take three men to turn it into a lifesaver.

There Is No Question, We Will Have to Go for Penicillin

When Chain met with Florey to go over his scholarly reading, they quickly agreed that their next research project should be a thorough examination of bacterial antagonists (growth inhibitors). They eventually chose three specific candidates for study: Bacillus subtilis (a species of bacteria found in dirt that doesn't cause disease but can promote healing), *Bacillus Pyocyaneus* (now *Pseudomonas pyocyanea*—a bacterium that turns pus blue and can inhibit the growth of other bacteria) and penicillin. But which to study first?

Penicillin's potency against staphylococcus appealed to Florey. "There is no question, we will have to go for penicillin," Florey told his friend R. Douglas Wright, an Australian physiologist whom he had brought to Oxford on a fellowship. "My worry is that... I've got my team together. If the money doesn't come along, I might not be able to hold them together and it would all be finished."

Chain began with the hypothesis that penicillin was an enzyme. Immediately, he ran into the problem that it was hard to grow enough of the mold to start a thorough investigation. He would grow the greenish-blue mold, wait for gold droplets of potent penicillin-containing broth to form on the dry fronds, and then draw them off with a pipette. They tested it on several bacteria and a coworker remembered, "The results were not impressive." It took ten days to produce enough of the juice to run one experiment. Then they would have to start the process again. The pace of progress was glacial.

Meanwhile Heatley, whose three-year grant was about to expire, had obtained a Rockefeller traveling grant to study in Copenhagen. But on September 1, 1939, the Nazis invaded Poland, and two days later Britain and their allies declared war against Germany. Suddenly, moving to the continent didn't seem wise. Florey seized the opportunity to ask Heatley to rejoin the team to work on penicillin, but Heatley flatly refused, knowing how unhappy he would be working under Chain. By now, however, Florey had become an expert at managing discordant personalities, and he quickly countered that Heatley could report to him instead of Chain. Heatley accepted and went to work, and Florey never bothered to divulge the new setup to Chain. Because Florey was present each day, ready to discuss the research, Chain failed to notice that Heatley was no longer his assistant. It wasn't until thirty years later that Chain learned the truth, admitting, "I understand now many actions of Heatley which were inexplicable to me at the time."

The impending war affected Florey and his coworkers in other ways as well. Florey wisely preempted the scattering of his team by carving out a new war-related role for the laboratory—they would work on blood transfusions, with Ethel heading a team of doctors that gathered blood from donors. Then, in the summer of 1939, every able-bodied man at the school helped to dig an air-raid shelter behind the building (even 5-year-old Charles, Florey's child, helped).

Research funds, which Florey had become adept at securing, dried up almost completely. He turned to the Rockefeller Foundation, writing, "Hitherto, the work on penicillin has been carried out with very crude preparations and no attempt has been made to purify it. In our opinion, the purification of penicillin can be carried out easily and rapidly. In view of the possible great importance of the above-mentioned bactericidal agents it is proposed to prepare these substances in a purified form suitable for intravenous injections and to study their antiseptic action in vivo."

Ordinarily Rockefeller officers' correspondence was conservative and muted, but Harry Miller, the foundation's man in England, wrote that "Florey practically is the only experimental pathologist of any real distinction in the British Isles. He is a distinguished young investigator who has the full confidence of Mellanby and I feel that his request if and when received should be given serious study." Within three months, the Foundation granted Florey more than he asked for £1,250 ($5,000) a year, guaranteed not just for the three years he requested, but for five years. As the months passed and the Nazi threat kept growing, the Floreys evacuated their two children to Cornwall, on England's west coast for safety, and they opened up their home to evacuees from London.

But Florey had his funding and set to work, breaking the problem of penicillin down into five puzzles: What is the best way to grow the most potent mold rapidly? Which bacteria does penicillin kill? How does it kill? What are the side effects on human cells? And what is its chemical structure?

Florey had found that his lab was most productive if he did not have committee meetings. Instead, each day he would make the rounds, talking to each researcher. Only when there were problems would he gather a few researchers to brainstorm. Heatley was happy reporting to Florey, and later explained, "Florey was in many ways the antithesis of Chain. He listened carefully to the views of the least experienced of his research staff, such as myself, and one could have a thoughtful and productive two-way talk with him. On occasion this could be intensely stimulating. One quickly learned that his advice was perceptive and often unexpectedly valuable, being presented in some modest form such as 'You might consider...,' or 'I doubt if you are right there, but why not try...'"

Florey had already set Heatley upon the task of obtaining sufficient quantities of penicillin to work with, and he proved absolutely the right man for the job. Gwyn Macfarlane said of him, "He was a most versatile, ingenious and skilled laboratory engineer on any scale, large or minute. To his training in biology and biochemistry he could add the technical skills of optics, glass and metal working, plumbing, carpentry, and as much electrical work as was needed....Above all, he could improvise— making use of the most unlikely bits of laboratory or household equipment to do a job with the least possible waste of time."

Heatley needed to scale up the process dramatically. The mold seemed to grow best on the surface of a solution no more than 1.5 centimeters deep, so Heatley grew it in anything shallow he could find, including trays, dishes, flat bottles turned on their side, and sixteen hospital bedpans borrowed from the infirmary. He added nitrates, sodium, salts, sugars, phosphate, glycerol, oxygen, and carbon dioxide—anything he could think of. None of these helped the mold grow. Around Christmas, 1939, a visiting friend of Florey's suggested adding brewer's yeast, which cut the fermentation time in half, although it did not increase the yield.

Over the next year, Heatley made improvement after improvement to the lab's brewing methods. He found that it was best to begin by sterilizing the containers in an autoclave. Then he would incubate the penicillin mold on a layer of Czapek-Dox liquid, a nutrient solution used to grow fungi. In a few days, a yellowish gelatinous film formed on top of the liquid, which became covered by green spores. Below it, the liquid turned yellow from dissolved drops of penicillin-containing broth, which would increase in quantity for ten days. Heatley found that if he drew off the nutrient liquid and replaced it, the mold would continue to produce more penicillin. He could usually do this up to twelve times before the mold stopped dropping penicillin into the liquid below.

Heatley also worked on making the mold extract more potent. First, he needed a reliable test that could compare its strength when he harvested it at different stages of growth. He began his assays using a technique of Florey's that involved a Petri dish with holes cut in it. He placed wax-like agar in the dish and colonized it with staph bacteria. Penicillin derived from varying stages of growth was dropped into the holes to see which had the greatest effect on the bacteria. This wasn't precise enough, so Heatley designed a technique that replaced the holes with porcelain tubes. This allowed for identical amounts of penicillin to be uniformly dropped into the bacterial colonies. After placing the penicillin into the tubes and allowing time to pass, he could measure the size of the growth-free halo around each cylinder, which indicated how much bacteria had been killed.

Meanwhile, Chain tested his initial hypothesis that penicillin was an enzyme, by putting it through a cellophane filter. Penicillin went right through it, proving it to be much too small to be an enzyme, and Chain had to admit that his "beautiful working hypothesis dissolved into thin air." But rather than becoming discouraged, he became fascinated: "It became very interesting to find out which structural features were responsible for penicillin's instability. It was clear that we were dealing with a chemically very unusual substance."

Chain tested penicillin's stability in solutions at varying pH levels and found that it was stable only at the edges of acidity and alkalinity, from pH5 to pH8 (pure water is neutral pH7, lower pHs are acidic, and higher are alkaline). Since he knew his penicillin broths contained many impurities, he tried conventional purification methods such as extracting the active substance by dissolving it in various solvents. The idea was to find a solvent in which penicillin was more soluble than it was in the moldy broth. Chain found that the weakly acidic penicillin could be extracted into ether, leaving the neutral impurities behind in the water layer. But when Chain tried to remove the penicillin from the ether, it seemed to vanish.

Over time, Heatley mechanized and greatly improved the extraction of penicillin into ether, building an apparatus to separate penicillin from its broth at a much faster rate. Once the penicillin-containing solution deposited by the mold was filtered to remove bits of mold and other contaminants, the liquid was jet-sprayed down a long tube, while a stream of ether (later replaced by amyl acetate) flowed upwards past it in the same tube, so that the two liquids had a large surface area of contact. The penicillin dissolved in ether and rose to the top, while much of the rest of the original solution sank and was drawn off. But the problem of how to remove the penicillin from the ether remained.

In March of 1940, Florey held one of his informal brainstorming meetings with Heatley and Chain to discuss their inability to extract penicillin from ether. Hesitantly, because he thought it was so obvious, Heatley speculated that reversing Chain's steps might work. If the slightly acidic penicillin would dissolve in ether, perhaps a slightly alkaline solvent would pull it back out of the ether. Florey was intrigued, but Chain immediately dismissed the idea, telling Heatley, "If you think it will work, why don't you do it yourself? That will surely be the best and quickest way to show that you are wrong."

Florey told Heatley to go for it, and Heatley set about filtering the mold juice to remove the large particles of mold, then mixing it with ether as Chain had done, and letting the ether, which now contained penicillin, rise above the remnant juice. He then mixed the lighter ether and penicillin mixture with an alkaline solution of buffer salts dissolved in water. He tested it and the water layer now held the penicillin. What's more, it remained in the water even eleven days later at room temperature. Heatley had found a way to stabilize penicillin. This allowed the researchers to create a "standard solution" and define a "unit" (now sometimes called the "Florey unit" or "Oxford unit" of penicillin), as the online Drugstore Museum explains: equivalent to "the smallest amount, which when dissolved in 50 cc. of meat extract broth, completely inhibits the growth of the test strain of *Staphylococcus aureus.*"

To make a drug, the research team still needed to precipitate penicillin out of the water. Freeze drying had been developed in 1935 in Sweden and had revolutionized the preservation of unstable materials: The process involves first freezing a material, and then using drying techniques that remove water (now ice) by making it form a gas without going through a liquid state, where chemical reactions are likely to occur and affect the substance. This is usually done at low pressure to promote evaporation, but you can observe it in your freezer, as well, when frost forms on surfaces that were previously dry; this is water that evaporated from other parts of your freezer and recondensed. Chain set to work. "The result," he later wrote, was "a very nice brown powder which kept the activity of the medium undiminished, without any loss whatsoever."

After almost two years of work, the team had a real drug that could be tested. Biological tests were Florey's specialty, and he chose Dr. Margaret Jennings to help, along with his trusted assistant, James Kent. Jennings, in her 30s, was a teacher and histologist at the school, and had worked with Florey on many research projects, often cowriting his papers. Her husband, a scientist in another department at Oxford, also had collaborated with Florey on a research project.

Heatley's engineered contraptions did not produce nearly enough penicillin to test it with abandon, so Florey sought out the pharmacologist, J. H. Burn, to teach his team the most efficient method of injecting mice. Burn said, "I remember the day when he and Mrs. Jennings brought a sample which they wanted me to inject into a mouse's tail vein. But when I had done it and the mouse was unaffected, they did not explain what it was, or why they wanted it done." By now, Florey suspected penicillin was remarkable, and he had gone into top-secret mode.

They ran test after test on mice, rats, rabbits, and cats, injecting penicillin into their skin, their blood, their abdominal cavities, giving it to them orally and through stomach tubes—all the typical tests drugs go through. They found that penicillin remained active when injected into the blood, but that it did not enter the blood stream when delivered to the stomach either orally or by a tube. However, it did enter the bloodstream when it was injected into the lower intestine by a tube. Chemicals in the stomach evidently destroyed it.

Yet even an injection of penicillin never lasted long. Within an hour or two it was gone from the bloodstream. Was it being destroyed or excreted? The team discovered that the mice were eliminating it in their urine, so they learned to recycle it by collecting urine, extracting the penicillin, and then re-administering it.

They ran toxicity tests. One of the most fragile cells in the body is the leucocyte, a white blood cell that engulfs foreign bodies as part of the immune system. They tested penicillin on leucocyte cells in test tubes and were stunned to find it harmless even when concentrated to one part per 500. Many other human tissue cells were also shown to be unaffected by penicillin.

At the same time, Florey organized other workers to do a wide survey on penicillin's effects on harmful bacteria, to determine what species it inhibited. They were surprised that it killed a diverse group of bacterial pathogens, although at varying doses. What was the mechanism of its power? Tests showed that penicillin did not kill bacteria immediately on contact, meaning it was not an antiseptic, and they knew it was not an enzyme that dissolved them. Under a microscope, bacteria treated with penicillin became swollen and elongated, stopped dividing, and eventually burst or died. Penicillin's potency was all the more stunning when it was discovered that if it was diluted to one part per million it still killed some types of bacteria, which meant it was twenty times more effective than the latest sulfa drug.

It Looks Like a Miracle

Finally, Florey and his team were ready to end their preliminary tests and perform a test on a living animal. At 11 AM on Saturday, May 25, 1940, Florey and Kent injected eight white mice with a dose of virulent streptococci known to kill mice. At noon, they gave two mice an injection of 5 ml of penicillin, and two others 10 ml. The other four were controls, so they received none. That afternoon, Florey, Heatley, and Kent watched the mice. At around 6 PM, Florey sent Kent home, and then gave the first two mice another dose. By 6:30, mice one, two and three looked fine. Mouse four looked so-so. The controls looked sick. The scientists went home to eat. At 10 PM Florey returned to give mice one and two another injection. The mice acted about the same as at six o'clock, so he went home. Heatley returned around eleven and began marking down the times of death of the four control mice. The last one died at 3:28 AM. The next day Florey, Heatley, and Chain met to go over the results. The four mice that had received penicillin were still alive. Florey, ever known for his understated manner, called Margaret Jennings. "It looks like a miracle," he told her.

On Monday, Florey and Jennings repeated the experiment using ten mice. On Tuesday, they used sixteen mice, doubling the amount of lethal streptococci they gave each mouse, and varying the doses of penicillin. In the months ahead, they treated mice infected with different diseases, using fifty to seventy-five mice for each experiment. They learned that for some pathogens they had to keep penicillin in the blood for long periods of time; because penicillin never lasted long in the blood, Florey and his team decided an intravenous drip was the best way to administer it.

That summer of 1940 was a foreboding time in England, as the nation prepared for a German invasion. If the Germans landed, even Cornwall might not be safe. The Floreys joined other Oxford University parents and sent their two children, Charles and Paquita, to the United States on a refugee boat. Paquita remembers their parting: "Dad didn't like fuss or emotional bother. He just kissed us goodbye and told us to be good, but mother was terribly cut up and cried because she felt it was a dreadful thing for us to go alone." Ethel chose to stay behind, judging that Britain would need her doctor skills. As a precaution, the lab drew up plans to save penicillin in the event of a German invasion. They would destroy all the records, but Florey, Heatley, and some of the others smeared spores of the penicillin mold into their clothes, where it could remain dormant for years. If they escaped, they would wear the spores to safety.

The team's clinical tests continued until they were confident that penicillin was a potent drug. They wrote up the results and on August 24, 1940, the prestigious British medical journal The *Lancet* published a two-page article titled "Penicillin as a chemotherapeutic agent." Florey expected a great deal of interest from other researchers and the pharmaceutical industry, but in September the bombing known as the Blitz started, and the German air force struck London for fifty-seven nights in a row.

The reaction to the publication was less than muted: Nothing happened. Even when Florey visited pharmaceutical companies and touted penicillin's benefits, none were willing to make an investment in a radically new drug during a time of uncertainty caused by the war. So Florey appraised the situation. He noted that a human was 3,000 times the weight of a mouse, which meant that they would have to produce at least 500 liters a week of the penicillin juice to test it on humans. With Heatley, he set to work building a factory in the school. They designed containers in the shape of bedpans, and ordered 600 of them from a ceramics factory. Heatley's contraptions were placed in room after room. Soon pipes were snaking through halls, and apparatuses were pulsating that would make Rube Goldberg proud. They hired six "penicillin girls" to help, and production began on Christmas Eve, 1940.

Once in a Life

That January, Florey and his team were ready to test penicillin on humans. Consulting with Dr. Charles Fletcher, they realized that tests on human subjects involved serious ethical issues, since mice and humans do not always react to drugs in the same way. They decided to first test penicillin on a consenting terminal cancer patient, who would not benefit from the drug, but who, if harmed, would die shortly anyway. Elva Akers, fully informed, agreed to be the guinea pig. She said penicillin had a curious musty taste, and several hours later she developed chills and a mild temperature, which the researchers took to be signs of impurities.

Over the next few weeks, the researchers worked at further purifying penicillin. Edward Abraham, another of Florey's lab researchers, had made an important contribution. Using chromatography—the technique of isolating a substance by drawing a mixture through various layers that act similar to filters—he separated the brown substance into bands of powder, one of which was yellowish and 80 percent pure. It turned out that the penicillin the team had been using all this time was only 1 percent pure. In the eighteen months the team had been producing penicillin, they had produced a total of 4 million Florey units for all of their experiments and human trials. The ideal dose for a single patient, it would later be learned, turns out to be 4 million units each day of treatment.

On February 12, 1941, they went to a septic ward where Dr. Charles Fletcher was treating a patient who had been infected for five months after scratching his face in his rose garden. The patient was Albert Alexander, a police constable, and he had already lost an eye to the infection. His lungs and shoulder were also infected, and he had abscesses over much of his body. Heatley wrote in his diary that Alexander "was oozing pus everywhere." After eight doses of penicillin, his infections were dramatically reduced, his temperature was normal, and his appetite had returned. Five days later, his swelling had almost disappeared. Regrettably, Florey and his colleagues ran out of penicillin, and Alexander relapsed and died.

Clearly, Florey and his team didn't have enough penicillin to treat adult patients with full-blown infections, so they decided to concentrate treatment on children and localized infections. They watched a carbuncle—an abscess the width of a hand—melt away from a patient's back over days of treatment. As Fletcher bicycled patients' urine back to Florey's lab to be recycled, he realized how privileged he was to be taking part in such a monumental project. "It is difficult to convey the excitement of actually witnessing the amazing power of penicillin over infections for which there had previously been no effective treatment," he recalled. In Florey and Chain he saw "the intense joy of the scientist seeing that years of work had resulted in an opportunity to save lives." Later, when he asked Florey how it felt, Florey said, "This is the sort of thing that only happens to you once in a life."

A Carpetbag Salesman
Trying to Promote a Crazy Idea

As so many people do in such times, Florey now honestly reassessed his personal life. With the children gone, he and his wife found they had nothing in common. Their relationship might have been best summed up by Paquita, who said to one of her aunts: "How extraordinary it is that the two most difficult people in the world should have happened to marry each other!" Florey began an affair with his assistant, Margaret Jennings, who was also unhappily married and separated from her husband. They would remain involved for twenty-five years, marrying only after Ethel died. Their arrangement was not known to the public in detail until after Margaret Jennings' death in 1994.

Florey also faced the fact that the factory he and his colleagues had cobbled together could never produce enough penicillin to treat all the people who needed it. Without the British pharmaceutical industry's help, Florey could only turn to one place—the United States. After obtaining government approval, Florey chose Heatley, who knew how to ferment penicillin, to fly with him to New York, much to the chagrin of Chain. On June 27, 1941, they boarded a plane for Portugal, with Florey holding the package of penicillin in his lap. The flight itself was not a minor matter, as it involved a dangerous nighttime leg around the war in Europe to Lisbon, a pivotal neutral city, much like that portrayed in the movie *Casablanca*. Next was a longer, twenty-hour flight across the Atlantic to Bermuda in a Pan Am Clipper, a giant amphibious aircraft, and then a final flight to New York.

The very day the scientists landed, they met with Dr. Alan Gregg, head of the Rockefeller Foundation's Medical Sciences Division, and without notes Florey told the story of the development of penicillin and its promise as a drug. Heatley remembered Florey's talk: "It was unemotional, but still very telling, and in a startling sort of way it revealed the wide grasp of his scientific mind. Even though I knew the subject well, he showed me new facts, and I realized suddenly how great a man he was.... I count that hour in Gregg's office as one of the great experiences of my life."

Immediately after the meeting, Florey drove to New Haven, Connecticut, to see his children, Paquita and Charles, and wrote to Ethel about them. Dr. Gregg was busy opening doors and after a flurry of meetings the researchers met Dr. Percy Wells, the head of the Bureau of Agricultural Chemistry and Engineering, a federal agency of the Department of Agriculture. Florey and Heatley expressed to Wells that great progress would have to be made in optimizing penicillin's growth if they were ever going to produce enough of the drug to distribute widely. Dr. Wells understood exactly what was needed and made arrangements for work on the mold to begin immediately at the Bureau's laboratory in Peoria, Illinois, which was already working on fermentation techniques for other uses. Florey and Heatley took their precious vials of penicillin by train to Peoria, and decided that Heatley should stay so that he could demonstrate all he knew about how to grow it. He settled in to work with the bureau's scientists for what would turn out to be a full year.

After such a fast start, Florey felt optimistic. Month after month though, one pharmaceutical company after another balked at the difficulties involved in making the drug. Not only would it require huge rooms of vats to grow enough mold to produce penicillin in great quantities, but even if it were as successful a drug as Florey promised, the companies were afraid other scientists would soon synthesize it, making their investment in fermentation technology immediately obsolete. Florey was left feeling like "a carpetbag salesman trying to promote a crazy idea for some ulterior motive."

However, Alfred Richards, the man with whom Florey had studied in Philadelphia fifteen years earlier, was impressed when he heard Florey's presentation. "Florey is a scientist and a scientist like that doesn't tell a lie," he said. By this time, Richards was the vice president of the University of Pennsylvania Medical School and chairman of the Committee on Medical Research and Development, part of the growing government military-scientific establishment. He started pulling strings, letting companies know that the government wanted penicillin produced. Eventually, Merck, Charles Frosst, Squibb, and Pfizer agreed to produce it.

While the companies began gearing up, the scientists back in Peoria had quick success in using corn steep liquor instead of yeast in the fermentation process. Corn wasn't grown commercially in England, but in Peoria its derivative was the first choice for all fermentations, for it was rich in nitrogen, which promotes growth of molds and other forms of life. The researchers realized that penicillin was produced by many different strains of the penicillium mold, so they began looking for ones that might produce a more potent extract. They had samples parceled to them from all over the world, but also sent a lab worker, Mary Hunt, to local supermarkets to bring back moldy fruit and vegetables, which they would then culture. One day "Moldy Mary," as she was known, brought back a cantaloupe infested with mold so powerful that it became the source of most of the world's penicillin for years. Now there were two penicillins, with the much more potent American version, Penicillin G, differing molecularly from the British Penicillin F. With the resources the major pharmaceutical companies brought to bear, many more incremental improvements would be made over the next two years.

It Was a Revolution

Florey returned to England and kept his lab working to produce penicillin, obtaining help from two other companies. In January of 1942, Ethel, also enthused, joined the team to conduct more clinical trials. These were great days for both Florey and his wife. He wrote to the now retired Sherrington: "The penicillin work is moving along and we now have a fairly substantial plant for making it here. It is most tantalizing really, as there is, for me, no doubt that we have a most potent weapon against all common sepsis. My wife is doing the clinical work and is getting astonishing results—almost miraculous some of them.... I am afraid the synthesis of the substance is rather distant, but if say, the price of two bombers and some energy was sunk into the project we could really get enough to do a considerable amount."

Later, Florey went to North Africa to oversee tests of penicillin on wounded soldiers. Most doctors were leery to give up their sulfa drugs, but penicillin's effectiveness quickly dispelled their reluctance. U.S. companies finally began producing enough penicillin for its first mass use, which was, oddly enough, to treat venereal diseases picked up by soldiers and sailors in the whorehouses of the Philippines and North Africa. One shot completely cured a soldier if he had been infected by syphilis for less than a year, and still does to this day (it takes three shots for those who have been infected longer than a year). By D-day, June 6, 1944, there was enough penicillin to treat all 40,000 men wounded in the Normandy invasion. The first British soldier to be treated was James Hill, commander of the 3rd Parachute Brigade, who had a buttock blown off by a German bomb. By the time the war ended, U.S. companies were making enough penicillin to treat a quarter of a million individuals a year.

Probably no other drug has ever had such a dramatic impact on medicine. Lewis Thomas, one time dean of Yale Medical School, described the impact on his generation of doctors: "We could hardly believe our eyes on seeing that bacteria could be killed off without at the same time killing the patient. It was not just amazement, it was a revolution." The mortality of young people with bacterial pneumonia fell from 66 percent to 6 percent.

Penicillin was magical like no other drug because it treated so many diseases: syphilis, gonorrhea, childbed fever, septicemia, meningitis, scarlet fever, gas gangrene, anthrax, tetanus, rheumatic fever, lobar pneumonia, and diphtheria. Its strength and broad spectrum gave a huge dose of giddy optimism to the medical community. A post-war account by well-known venereologist George Bankoff stated: "Soon young ladies will be able to buy their lipsticks impregnated with penicillin. They will still have their lips made beautiful and inviting, but the danger of infection that every kiss potentially can transmit will be removed. Penicillin will be like a guardian angel ready to halt any intruder that is unwelcome. Facial creams, mascaras and of course all toothpaste will be impregnated with penicillin too, thus preserving skin and teeth healthy and unblemished."

It is an interesting fact of human nature that we count deaths, but not lives saved. Almost all of us have been dosed with penicillin or its derivatives at one time or another, sparing us pain and suffering, and in many cases death. Amy Pearce examined the best available data, looking at mortality rates before penicillin existed and after it was available, and her estimate of the number of lives penicillin has saved is staggering. Millions of children have been saved from pneumonia and millions of adults' lives have been saved from syphilis in the United States and Europe alone. Tens of millions have been saved from meningitis and blood poisoning. There is scant data from the third world, although penicillin is widely used in all countries, and is often available over the counter. But counting only the developed world and these four diseases, the number of lives saved is over 80 million.

Once penicillin gained publicity, thousands of scientists began studying it. In 1945, Dorothy Crowfoot Hodgkin of Oxford University determined penicillin's structure using X-ray crystallography and went on to win a Nobel Prize for her work on it and other substances, including insulin. It was found that penicillin works by binding to an enzyme that many bacteria use to build their cell walls. Without this enzyme, the walls of these bacterial cells weaken and collapse.

Penicillin itself has also undergone vast development since Florey's day. In order to slow down the rapid excretion of penicillin from the body, a search began for a molecule that would compete with penicillin in attaching to the organic acid transporters in the blood that were ushering penicillin so quickly from the body. It was found that administering probenecid—a drug used to treat other ailments, including gout—allowed penicillin to last much longer in the body, all the while killing bacteria. In 1957, John C. Sheehan of MIT synthesized penicillin, rendering unnecessary the fermentation process Heatley so labored over. And, in the 1960s, the Beecham Research Laboratories in Surrey, England, patented the synthesized penicillin derivatives ampicillin and amoxicillin, which made good absorption of oral penicillin possible in pill form. Fifty years after the advent of penicillin, amoxicillin was still the most prescribed drug in the United States, constituting 3.8 percent of all the prescriptions in the country in 1994.

The popularity of penicillin also led to a broad search for other natural substances with antibiotic properties. In 1944, Selman Waksman, the man who coined the term antibiotic, discovered streptomycin, which treated tuberculosis, long a scourge of humankind. Abraham and Heatley, still working under Florey, went on to develop the cephalosporins, a class of antibiotics that kill some penicillin-resistant bacteria. Today, more than 10,000 substances with antibiotic properties are known, many of which have been used to save millions of additional lives.

The Problem of Resistance

Four years after the mass-production of penicillin began in 1943, resistant strains of *Staphylococcus aureus* bacteria emerged. Staph a. is a normally benign organism that many of us carry on our skin and in our noses. When it invades our bodies it sometimes causes minor infections such as boils and urinary tract infections, but it can also cause serious ones such as meningitis, pneumonia, food poisoning, and toxic shock syndrome. It was this drug resistant staph that gave Muppets creator Jim Henson pneumonia, from which he died.

Bacterial drug resistance occurs by evolution through mutation. Individual bacteria with mutations that allow them to survive an antibiotic thrive and replace those without resistance. For instance, some populations of staph a. were observed in 1947 producing an enzyme that would break apart the beta-lactam ring in penicillin, which is the very structure that gives penicillin its antibiotic potency. Bacterial evolution can occur rapidly. Bacteria (and many viruses) can reproduce as fast as every twenty minutes: This equals three times an hour, seventy-two times a day, 26,280 times a year, and 525,600 times faster than a human's normal reproductive cycle of around twenty years. In addition, an infection consists of many bacteria. When an antibiotic is taken, it may kill billions of bacteria, but if only a few with mutations survive, grow, and multiply, they will quickly repopulate, carrying with them their resistant characteristic.

Studies of bacterial resistance demonstrated that not only can bacteria pass mutated genes along to their offspring by asexual reproduction, but also through horizontal transfer, known as bacterial conjugation. Once a mutated gene exists in a bacterial population, it can be passed to its neighbors by the transfer of plasmids, which are strands of DNA independent from the chromosomal DNA. Soon a whole colony of bacteria can have its own life-saving gene which can be passed on to their progeny.

Scientists have worked diligently to develop new antibiotics to replace those that become obsolete due to evolved resistance. But by the late 1990s, half of all Staph a. bacteria was resistant to penicillin, methicillin, tetracycline, and erythromycin. Two million people in the U.S. become infected in hospitals each year, according to the National Institutes of Health. Seventy percent of the bacteria that cause these infections are resistant to at least one antibiotic. Of those infected, 90,000 die, an increase of 9 percent in the past ten years. Bacteria are as clever at developing resistance to current antibiotics as scientists are at developing new ones, and the battle against infection, begun in 1942 with the development of penicillin, is far from over.

The Fleming Myth

When news of penicillin broke, the media turned their attention to the discoverers. When reporters first showed up in Oxford, however, Florey literally ran out the back door. He refused to play the celebrity game, and he refused interviews, well aware that there was still not enough penicillin to treat all who needed it all over the world. He explained, "It is utterly wrong to write about this drug as 'magic.' It is not a cure-all, and it is cruel to raise hope among the dying and their relatives that such a substance exists and then to tell them that they cannot have it supplied."

Alexander Fleming, in contrast, relished the spotlight and granted interview after interview. The press ignored the fact that penicillin had lain fallow for a decade before Florey's team picked it up. The reporters knew nothing of the years of chemical, engineering, and clinical work that was necessary to turn a mold into the first antibiotic drug.

This lack of recognition culminated in a letter Florey received from St. Mary's Hospital. It started, "You may have heard of the discovery of penicillin by Professor Fleming at St. Mary's Hospital," and invited him to buy tickets for a charity show. Florey laughed, then had it framed and hung on the wall at the entrance of the Dunn Laboratories. Others at the lab didn't appreciate the irony, and tried to convince Florey to set the record straight with the press, so he turned to Mellanby and Sir Henry Dale, the President of the Royal Society. Both urged him to maintain a British stiff upper lip, and keep quiet. As he followed their directives, the press concentrated on a willing Alexander Fleming. The Oxford team, whose diligent work had actually created the drug, became a forgotten footnote in the public story that became notorious as the Fleming Myth (even Fleming called it that), which Fleming greatly enjoyed. Although he was always careful to give credit to Florey's team, Fleming was lionized, and relished it. Indeed,

Florey's team's contribution was assumed to be minor, while Fleming became one of the most famous men in Europe, traveling all over, posing for pictures, and lecturing. His name still comes up on lists of the greatest scientists of the twentieth century.

In 1945, Howard Florey, Ernst Chain, and Alexander Fleming received the Nobel Prize for Physiology and Medicine. Fleming had discovered penicillium—the mold; Florey and Chain and Heatley had discovered penicillin—the drug. The Nobel committee limits its awards to three people, and as a result, Norman Heatley was overlooked in the committee's decision. The omission of Heatley was such an injustice that in 1990, to mark the 50-year anniversary of penicillin as a drug, Oxford University conferred upon the 78 year-old Heatley its first honorary doctorate of medicine in its 800 year history.

Working on Mucus

One day in the 1950s, a young student saw an older man in the hallway at Oxford and asked a friend who he was. His friend replied, "At breakfast a couple of days ago I saw that chap sitting on his own. He let me join him. I asked him what he was doing. He said he was working on mucus."

"My God," the student replied. "How sad. Here's a man probably in his 60s, probably at the end of his career, studying mucus."

Such were the accolades of the young. The man was Howard Florey, leader of the team that created penicillin, the most revolutionary drug of all time. His peers made him president of the Royal Society in 1960, the highest honor in British science, and in 1963 he became president of Queen's College, one of the oldest of the colleges that make up the University of Oxford. He worked until the day he died, February 21, 1968, at the age of 69.

Lives Saved: Over 80 Million

Discovery: Penicillin

Crucial Contributors

Howard Florey: Formed the team, initiated the research, and guided the development of penicillin along all steps until it became a revolutionary drug—the first antibiotic.

Ernst Chain: Contributing from the beginning, he developed procedures to purify penicillin and freeze-dry it.

Norman Heatley: Developed methods to grow the mold that produces penicillin, developed assay techniques, and developed an extraction routine.

Alexander Fleming: Published his observation that the mold had antibiotic properties.

In 1946, Howard Florey received the Albert Gold Medal. J. A. Hampton/Hulton Archive/Getty Images.

"There is no question, we will have to go for penicillin. My worry is that I've got the bacteriologists and biologists, and I've got my team together. If the money doesn't come along, I might not be able to hold them together and it would all be finished."
—Howard Florey

"I have never met an educated man whose conversation stayed so close to the ground. It was as if he deliberately shunned subtle or polished speech."
—Professor Henry Harris

"Nevertheless, it was Florey's personality that captured so many able young research workers, who had been initially attracted by his scientific reputation. What they found in him was an infectious vitality: great physical energy combined with an independence of mind that seemed to open mental windows and let the fresh air of realism clear away stuffy academic pomposities."
—Gwyn Macfarlane, associate and biographer

"One becomes rather lost in a maze at the thought of stopping the appalling thing of seeing young people maimed and wiped out while one can do nothing."
—Howard Florey

Houdini died in 1926 of an appendicitis infection that penicillin likely would have cured.

A 1944 bacterial culture being grown in a petri dish. Florey's team tested the strength of penicillin by dropping it into the porcelain tubes and measuring the circle of killed bacteria. Fritz Goro/Time & Life Pictures/Getty Images.

On each square centimeter of human skin reside about 100,000 bacteria. Living mainly in our gut and comprising over 500 species, bacteria add about three pounds to our body weight.

"It looks like a miracle."
—Howard Florey, upon seeing that mice infected with virulent streptococci survived when treated with penicillin.

Alexander Fleming gets most of the credit for discovering penicillin, but he only discovered penicillin: the mold. Florey's team discovered the much more difficult and important penicillin: the life-saving drug.

"We could hardly believe our eyes on seeing that bacteria could be killed off without at the same time killing the patient. It was not just amazement, it was a revolution."
—Dr. Lewis Thomas

Until penicillin, as Charles Fletcher, the doctor who ran the first tests on penicillin, describes it, "every hospital had a septic ward, filled with patients with chronic discharging abscesses, sinuses, septic joints and sometimes meningitis…chambers of horrors, seems the best way to describe those old septic wards." Carbuncles, abscesses the size of one's hand, oozing pus were not uncommon. There was little treatment. About half the people who entered the wards exited on gurneys to the morgue.

A man working in a 1944 penicillin plant. Producing penicillin required a huge industrial investment. Fritz Goro/Time Life Pictures/ Getty Images.

In the eighteen months Florey's team had been producing penicillin, they had produced a total of 4 million units for all of their experiments and human trials. It would be learned that is the amount of one single daily dose of treatment.

When Florey tried to convince skeptical pharmaceutical companies to make penicillin he said that he felt like "a carpet bag salesman trying to promote a crazy idea for some ulterior motive."

Penicillin was magical like no other drug because it treated so many diseases—syphilis, gonorrhea, childbed fever, septicemia, meningitis, scarlet fever, gas gangrene, anthrax, tetanus, rheumatic fever, lobar pneumonia, and diphtheria.

With the use of penicillin, the mortality of young people with bacterial pneumonia fell from 66 percent to 6 percent.

Fifty years after the advent of penicillin, amoxicillin, in the penicillin group of drugs, was still the most prescribed drug in the United States, constituting 3.8 percent of all the prescriptions in the country in 1994.

Chapter 9: Frederick Banting— Over 16 Million Lives Saved

Insulin: The First True Miracle Drug

LEONARD THOMPSON, age 14, lay in the public charity ward of Toronto General Hospital in Canada in the final stages of consciousness before slipping into a coma. He was listless and pale, his hair falling out and his abdomen distended. Every time he breathed, the room filled with the sickly aroma of rotten fruit—the stench of the final stages of his disease. He weighed just 65 pounds on that day in January 1922, and he had, at the most, just weeks to live.

In Washington, D.C., the 14-year-old daughter of the Secretary of State gazed at the hair covering her pillow. Elizabeth Hughes' hair had been falling out by the handfuls for weeks as her body desperately sought to conserve every bit of energy she was allotted on her starvation diet of 700 calories a day. She weighed 44 pounds and stood barely five feet tall, resembling a tiny skeleton more than a teenaged girl. She only remained alive, her mother knew, out of sheer determination.

And in Rochester, NY, Jimmy Havens, 22, weighing 73 pounds, spent his day crying from pain, hunger and despair. "Please let me die," he pleaded with his family. They couldn't. They just loved him too much. All they could do was watch helplessly as the disease destroying their son and the starvation diet "treatment" competed to see which would kill him first.

All suffered from a disease that, in 1922, was a death sentence: diabetes. With no cure available, each would die within a few months unless a miracle occurred.

Average Student and Poor Speller

Frederick Banting was born on November 14, 1891, on his family's farm in a small county in Ontario, Canada. He was the youngest of five children in a family that prayed as hard as it worked. Banting really did walk five miles to school every day, through the worst a Canadian winter had to offer. A shy boy, he hated school, where he was often teased mercilessly. He was, at best, a mediocre student, too afraid of getting the answer wrong to raise his hand, too ashamed of his copy books filled with spelling errors to show them off. Since his brothers were much older than he, much of his free time was spent on the farm, playing with the animals he loved and staying close to his favorite person in the family, his mother.

Although few farm boys at that time managed to finish eighth grade, let alone high school, Banting knew he wanted to attend university. He just didn't know why. Maybe it had something to do with his love of books. Every evening he would lie on the floor of his room above their small kitchen, ear pressed to the hole in the floor that let in heat from the kitchen stove, as his father read novels by Walter Scott and Charles Dickens, or poetry by William Wordsworth. He loved the tales of adventure and hardship, of the worlds those writers created so far away from the farm. Someday, he vowed, he would leave the farm and find that world. College offered a way to do that.

He barely passed his high school exams, and his high school principal later related, "We would not have picked him for one on whom fame should settle." Banting still managed to get accepted into Toronto's Victoria College in 1910. His goal—or, rather, his father's goal for him—was to become a minister. By then Banting was nearly six feet tall and big-boned, with a frame that spoke of the athlete he'd been in high school. He was still a terrible speller, an affliction that would haunt him throughout his life and be evident in all his letters and notes. He was also famously stubborn, and the few beatings he received in his childhood were for an intractable volition that his parents hoped would, at worst, develop into doggedness.

Yet Banting wasn't sure he felt called to the ministry. Since childhood, he had always had a fascination with medicine. As a young boy, he had seen two construction workers fall off their scaffolding and suffer severe injuries. Young Frederick ran for help and then watched, mesmerized, as the doctor examined and treated the stricken workers. "In those tense minutes, I thought that the greatest service in life is that of the medical profession," he later wrote. "From that day it was my greatest ambition to be a doctor." Banting dropped out of college in his second year and spent that spring and summer working on the farm with his father. While it took a lot of talk, prayer, and convincing, his parents finally agreed to help pay for his medical education. In the fall of 1912, Banting enrolled in the University of Toronto's five-year medical program.

Mustard Plasters and Fee Collections

Banting began his career during a time of great change in the medical profession. Prior to the twentieth century, most practicing North American doctors obtained their education through an apprenticeship program, spending a few months a year for two years at a physician-owned "medical school." If they could afford it, they went off to Germany or Austria to train. As late as the 1870s, most medical schools in Canada and the U.S. were without laboratories. Few, if any, had any admissions requirements, and anyone who wanted could be admitted to medical school.

By the 1890s, however, a new model of medical education had emerged, one which encompassed hands-on laboratory work and substantially more instruction in the basic and medical sciences during the preclinical years. This shift took off in 1910 with the publication of the Flexner report, sponsored by the fledgling Rockefeller Foundation, which showed the sad state of most medical schools in North America, and which led to the closing of all but a handful. That same year, the University of Toronto moved its medical education into the modern era.

At the turn of the twentieth century, the enormous possibilities of modern medicine were only beginning to become evident. Aspirin, the first synthetic drug, came on the market in 1899. Created and produced by the German company Bayer, it marked the birth of the pharmaceutical industry. Beyond that, however, doctors had few cures or even working treatments for most diseases. Millions of children died before their 10th birthday of common infectious diseases like measles, typhoid, and diphtheria (unfortunately this is still true today in parts of the developing world). Vaccines were rare, and in those days before antibiotics a simple infection could be a death warrant. Doctors used a few rudimentary procedures to treat everything from fever to cholera. Blood letting was popular, in which to remove impurities doctors opened veins and "bled" patients, or placed leeches on them to suck out their blood; and poultices such as mustard plasters were used to "sweat out" illness-causing impurities. Small wonder patients often died from the treatment instead of the disease.

Banting received training in all these approaches, as well as successful doctor tips, such as when confronted with a syphilitic patient to "tell them what your fees are and collect them first." He also spent a great deal of time in the operating rooms and soon developed a love of surgery. But it wasn't until his fourth year, while attending a lecture by the leading diabetologist Frederick Allen, that he learned about diabetes and its only treatment: starvation. The lecture made little impact on him, however, and his notes from that day say only: "—not dangerous to starve patients—not always best for them to maintain their own weight."

He graduated with his medical degree on December 9, 1916. Canada was deeply involved in World War I, and like so many of his classmates, he reported for military duty the following day. After training, he was assigned to the same orthopedic surgeon under whom he'd trained in Toronto, and by January 1917 he and the surgeon shipped out to Kent, England to treat patients injured during battle. There, he wrote a former girlfriend, he assisted in surgery three times a week, overseeing the care of more than 100 patients while also playing doctor to several Canadian families. Like many soldiers he also found time to do a bit of carousing. For Fred Banting was, as his biographer Michael Bliss described him, "a man's man." He liked to drink and he liked the ladies.

That ended in June 1918, when Banting was transferred to the front in France and saw real battlefield medicine. His job was to dress and clean soldiers' wounds before sending the often-dying men on to hospitals in the rear. Then came September 27, 1918. Caught in a terrifying battle with the Germans, forced to dive into ditches for cover from exploding shells, yet determined to continue helping with the wounded, Banting tried to make a dash for the medical tent. A piece of shrapnel shredded his right arm. Despite heavy bleeding and pain, he tried to keep working with the wounded until his commanding officer pulled him out and ordered him to get help. His courage under fire later won him the Military Cross, with the official citation noting that "his energy and pluck were of a very high order."

But his military career was over. For weeks, his wounded arm failed to heal. At one point, doctors wanted to cut it off, but Banting wouldn't allow it—to do so would have meant the end of his surgical career. Instead, he turned his care over to the doctor he knew best—himself. After months of convalescence, by February 1919, he was back in Canada with his arm intact and functioning. He spent six months at a Toronto military hospital helping set and mend the broken bones of soldiers before receiving his formal discharge at the end of the summer of 1919. He was 28-years old, a smoker and heavy drinker, and was engaged to a high school teacher named Edith. But he really had no idea what to do next. Like so many of that post-war generation, he went searching.

Banting decided to try a year at Toronto's Hospital for Sick Children, and in the next six months he assisted with 100 surgeries and performed an additional 100 on his own, but he wasn't offered a permanent position, and in the spring of 1920 he found himself back where he'd started, trying to figure out what to do with the rest of his life. Restless, and not willing to marry without a source of income, he decided to set up his own medical practice. Edith lived in London, Ontario, a city of 60,000 about 120 miles west of Toronto. Banting knew it was not overrun with surgeons as was Toronto, so, with a loan from his father, he bought a large brick house for $7,800 and hung out his shingle, using part of the downstairs for his office.

He waited for patients. And waited. And waited.

It took three weeks before he saw his first patient, and by the end of his first month he had earned a grand total of $4. After a couple of months, he was so bored and broke that he took a job teaching surgery and anatomy to medical students at London, Ontario's Western University. On October 31, 1920, preparing for a talk on diabetes and carbohydrate metabolism (about which he knew only what he had learned at the one lecture in medical school), he picked up the November issue of *Surgery, Gynecology and Obstetrics*, which had arrived the preceding day. He intended to use the journal's dullness to soothe him to sleep. Instead, he found himself fascinated by the lead article, "The Relation of the Islets of Langerhans to Diabetes with Special Reference to Cases of Pancreatic Lithiasis," written by Moses Barron. That one article, innocuous as it sounds, set in motion a medical revolution.

Sweet Urine and Death

Diabetes has likely been around as long as mankind. The early Greeks recognized the disease and bestowed upon it the name diabetes, which stems from the Greek word for "siphon," or "pipe-like." This refers to the tremendous amount of urine that diabetics produce. Early accounts tell of flies and wasps hovering over chamber pots, drawn there by the high sugar content of the urine. In 1675, the British physician Thomas Willis tasted the urine of several diabetics and found it uniformly sweet, and for nearly two centuries after, tasting urine was the primary test for a diabetes diagnosis. Today, we know the reason diabetics' urine is so sweet is that their bodies are unable to process glucose, the sugar molecule that provides the energy that cells need to power every aspect of human life.

Even in Banting's time doctors recognized that there were two types of diabetes: a type they called "juvenile" or "childhood" diabetes, because it mainly occurred in children, and one called "adult" diabetes, which usually developed in mid-life or later. The childhood version was deadly, striking suddenly with symptoms of great thirst and urine output, tremendous hunger, and frightening weight loss. Known today as type one diabetes, it affects some 2 million people in Europe and North America and can also strike adults. We also know that it is an autoimmune disease, triggered when something—such as a reaction to a virus—flips a pre-existing switch in an individual's genetic code that directs the immune system to attack the pancreatic cells that make insulin. Insulin is the hormone that unlocks muscle and fat cells to let in glucose. When insulin binds to receptors on the cell membrane known as glucose carrier proteins, a chain of events is set in motion that culminates in the cell synthesizing proteins that allow glucose to cross the membrane. Without insulin, these cells can't get access to glucose to make energy. Without energy, they can't carry out the functions vital for all animal life.

The 100 percent mortality rate of type one diabetes defied hundreds of years of efforts to find a way to arrest this dreadful disease. At first, doctors prescribed great quantities of sugar to replace that lost in the urine of diabetic patients, but this only hastened their deaths. In 1776, the discovery of sugar in the blood of patients confirmed that diabetes was a systemic disease, not simply one of the urinary tract. A few years later, the researcher Thomas Cawley performed an autopsy on a patient who had died from diabetes and found the pancreas, a small, jelly-like gland that sits atop the liver, shriveled up, the first intimation that the gland plays a role in the disease. But it wasn't until 1889, twenty years after the German scientists Joseph von Merring and Oskar Minkowski first hypothesized that something within the pancreas was responsible for the terrible sugar disease, that Minkowski removed a dog's pancreas and observed the previously housebroken dog suddenly peeing everywhere. When he tested the dog's urine, he found it filled with sugar. Minkowski had become the first person to induce diabetes in another living being. Even more importantly, Minkowski found that when he cut the pancreatic duct leading to the dog's intestine, thus preventing release of the pancreatic enzymes known to assist in digestion, he didn't induce diabetes. Obviously, some other part of the pancreas was at work.

That other part had first been described in 1869 by German medical student Paul Langerhans. Langerhans found that the pancreas contained two types of cells: the acini, which are clusters of cells that secrete the pancreatic enzymes, and another type unconnected to the acini. A few years later, the French researcher Laguesse bestowed the name "islets of Langerhans" on those other cells—which looked like islands floating in the pancreas—in honor of the man who originally identified them. There are from one to three million of these islet cells in a normal pancreas, constituting about 2 percent of its mass. In 1893, the French researcher E. Hédon removed nearly the entire pancreas from a dog, including the part that secreted the digestive enzyme, leaving only a small portion with the islets of Langerhans. This he grafted under the dog's skin so that it continued functioning. The dog didn't develop diabetes. But when Hédon cut off the remaining pancreas. . . Voila! Diabetes. Clearly, something within the islets of Langerhans prevented diabetes. Over the next few years, more than 400 researchers set to work trying to find that mysterious "something." As diabetes researcher Lydia Dewitt estimated at the turn of the century: "More thought and investigation was going into the islets of Langerhans than any other organ or tissue of the body."

On June 20, 1906, a German internist named Georg Ludwig Zuelzer injected an extract of cow pancreas and adrenalin into a man in a diabetic coma. The man came out of the coma feeling hungry, but he died twelve days later when the extract ran out. As Zuelzer later wrote:

"Whoever has seen how a patient lying in agony soon recovers from certain death and is restored to actual health will never forget it." Zuelzer's was just the first of what would be many such accounts on the miraculous recovery the right extract could bring to diabetics on the brink of death. But the recovery never lasted and inevitably the patients had severe reactions, including vomiting, high fevers, and convulsions. By 1913, diabetologist Frederick Allen declared: "All authorities are agreed upon the failure of pancreatic opotherapy [pancreatic juice therapy] in diabetes . . . injections of pancreatic preparations have proved both useless and harmful."

By the time Banting read the diabetes article on that Halloween night, the two primary treatments for diabetes were opium, to make the patient oblivious to the pain, and the infamous "Allen diet." This diet called for starving patients until their urine contained no sugar, then gradually reintroducing food until the ideal amount of calories to sustain the sugarless state was found. Victims often lived on rations of 400 to 600 calories a day (which you can get in a single large order of fries from McDonalds), becoming so weak they couldn't leave their beds, so desperate for food they stole birdseed from the family canary. Years later, when pictures of World War II concentration camp survivors circled the globe, doctors were reminded of those early diabetic patients. The estimated lifespan for someone who developed type one diabetes was a year to eighteen months without the Allen diet, four or five years with it. But the end was always the same—death.

Getting to Toronto

The article Banting read that October night didn't have its desired physiological effect; instead of falling asleep, he found himself wide awake. The idea outlined in the paper intrigued him. The author wrote that while conducting an autopsy on a patient he had found a small stone obstructing the main pancreatic duct, through which the strong pancreatic enzymes pass to the stomach to help digest food. The bulk of the pancreas had wasted away from the obstruction, but the cells in the islets of Langerhans were still healthy. Possibly, the author noted, by deliberately destroying most of the pancreas except the part containing the islets of Langerhans, one might be able to isolate that mysterious internal substance that played a role in diabetes. Find that substance, he wrote, and you might be able to cure diabetes. Could it be, Banting thought, that no one had been able to isolate the mysterious substance from the islets because it was destroyed by the potent pancreatic enzymes? Was there a way to isolate the islets from the rest of the pancreas?

He got up and reached for some paper and a pen to scribble down his thoughts: "Diabetus [note the misspelling]. Ligate [tie off] pancreatic ducts of dog. Keep dogs alive till acini degenerate, leaving islets. Try to isolate the internal secretion of these to relieve glycosurea [sugar in the urine]."

As later described in his obituary in *Time* magazine, Banting thought the mysterious substance secreted by these islets must act like a spark plug, providing the "juice" to help the body metabolize carbohydrates. He wasn't far off.

That next day, despite his sleepless night, Banting's enthusiasm remained at a fever pitch as he felt the excitement of the scientist who has just seen a novel approach to a problem. He brought his idea to the head of the physiology department, F. R. Miller, who suggested he visit John James Rickard Macleod, then a professor of physiology at the University of Toronto and one of the world's foremost experts on carbohydrate metabolism. The next week, while in Toronto for a wedding, Banting called on Macleod. Macleod, 44, was the Scottish son of a minister. A small, well-dressed, formal man, he'd become an authority on carbohydrate metabolism while at Western Reserve University in Cleveland, when he published a monograph in 1913 called "Diabetes: Its Pathological Physiology." But in 1914, he reached an impasse in his search for the cause of high blood sugar. Although he'd published work noting that there was probably some secretion from the pancreas involved in diabetes, he thought it would never be feasible to separate it out. By the time Banting came to see Macleod in November 1920, the professor's research focus had shifted to metabolism. He believed some part of the brain controlled glucose storage in the liver. The failure of this mysterious part of the nervous system, he and others speculated, was behind the high blood and urine sugars seen in diabetic patients.

The first meeting between the two men on November 7th foreshadowed their future relationship. Macleod was skeptical of Banting's idea. For his part, Banting was shy and bumbling in his presentation (indeed, he would never become a good public speaker, no matter how many opportunities he received). At one point, historian Michael Bliss writes, Macleod "began reading some of the letters on his desk, a sometimes unconscious and common gesture bound to offend the sensitive visitor to his office." When Banting didn't take the hint and leave, Macleod brusquely told him that some of the finest researchers in the world working in some of the finest laboratories available had failed to puzzle out the pancreatic extract issue, and thus it was doubtful Banting would be successful. "Are you sure you are prepared to undertake this, Dr. Banting?" Macleod asked, adding that it would require full-time focus and commitment. If ever there were a question guaranteed to activate Banting's doggedness, this was it. He assured Macleod he was committed. But Macleod knew that what Banting wanted to do wasn't completely unique. Several other researchers had tried the ligation approach without success. But none had tried to find the extract in a fully atrophied pancreas. "Maybe next summer," Macleod said, dismissing Banting.

Banting returned to London highly disappointed at being put off. He wanted to close up his practice and start the research immediately. But his fiancée, Edith, told him to forget about diabetes and to settle down and run his medical practice. Banting tried, but he could not stop thinking about diabetes. It seemed to consume him. What has never been clear is just why that article, resulting in Banting's first research hypothesis, so intrigued him. After all, research had never been a top priority to him. Maybe it was the surgery that would be involved; maybe it was a way to escape the boredom of his daily routine; or maybe he truly felt inspired. Regardless, he contacted Professor Macleod again in January and once again in March, and this time he put his thoughts into writing, a move that definitely improved his standing with Macleod, who still put him off until the coming summer.

Impatient by nature, Banting became discouraged after the March meeting. He learned of an opening for a doctor to accompany an expedition to the far north of Canada as its medical director. He later wrote: "I was in desperate circumstances. The possibilities of doing research seemed remote. I tossed a coin, three out of five. Heads, I was to do the research, tails I was to go to the Arctic to search for oil. Tails won, and I took the next train to St. Thomas to make a personal application in the hopes of obtaining the job."

A few weeks later, a letter arrived with news that the Arctic party had decided not to take a doctor. So, after giving final exams to his students in London that spring, Banting took the train to Toronto on May 14.

Dying Dogs and Isletin

When Banting arrived in Toronto, Macleod had left the country for the summer, but had ten dogs, a filthy laboratory on the top floor of the university's medical building, and a fourth-year student waiting for his attention. The student was Charlie Best. Macleod had chosen him and another student, Clark Noble, to help Banting through the summer. The idea was that they would each work a month and vacation a month. To see who would get to vacation first, they flipped a coin. Best lost. The first order of business was scrubbing down the filthy lab, which apparently hadn't been cleaned in years. Banting and Best washed the walls and floor on their hands and knees until the lab was sparkling. Then they turned to the dogs.

The first dog, number 385, was a brown female spaniel. Spaniels were (and still are) often used in medical research because they're small and relatively calm. The plan was to anesthetize the dog, cut her open, pull out her pancreas, tie off the pancreatic ducts, and close the incision. Ideally, then, the main part of the pancreas would shrivel up or atrophy, leaving the precious islets of Langerhans and their mysterious substance isolated. Other dogs would have their whole pancreas removed, giving them type one diabetes, so they would be available to test the mysterious substance once it was purified. Unfortunately, dog number 385 died the first day from an anesthesia overdose. By the end of the second week, seven of the ten dogs were dead. Because there was no budget for additional dogs, to continue their research Banting and Best took to buying dogs on the streets of Toronto with their own money.

It was hot that summer, especially in the sweltering top-floor laboratory with no air-conditioning, so they often took the dogs up to the roof to catch the breezes. After seven weeks and fourteen dead dogs, and still with no sign of pancreatic atrophy, Banting and Best realized their mistake. They were using catgut to tie off the pancreatic duct, and it loosened before any damage could set in. Banting sold his car and bought more dogs. This time, he used surgical silk for the ligatures, and by July 30 he had two atrophic pancreases. Banting and Best then made an extract from the desiccated pancreases and injected 4 ccs into a small white terrier with a blood sugar of 200 mg/dl (milligrams per deciliter). An hour later, the terrier's blood sugar had dropped 40 percent to 120 mg/dl. But despite more injections, the sugar levels began to rise again within a couple of hours. The next day, the dog died of diabetes. Another trial two days later also failed. Two attempts, two dead dogs.

They were running out of diabetic dogs.

On August 3, Banting did a total pancreatectomy on a collie. The following day, he and Best gave the collie some of the terrier's remaining extract. To their amazement and joy, the dog's blood sugar dropped with each injection. Banting and Best continued experimenting on the collie, injecting extracts of liver and spleen and a boiled version of the pancreatic extract, to no effect. Although the dog died on August 7 from a massive infection, Banting could at last write Macleod about their progress: "I have so much to tell you and ask you about that I scarcely know where to begin. I think you will be pleased when you see how the problem is unrolling from one end and rolling up at the other. At present I can honestly state my opinion that (1) the extract invariably causes a decrease in the percentage of blood sugar and in the excretion of sugar in diabetic dogs. (2) That it is active at least for four days if kept cold. (3) That it is destroyed by boiling. (4) That extracts of spleen and liver at least, prepared under similar conditions, have no such action. (5) The clinical aspect of the animals is improved by the extract. The number of problems that are presenting themselves is becoming greater and greater. Some of them I would wish to present for your approval...."

Excited and energized by their progress, they worked around-the-clock over the next two weeks, pancreatizing more dogs and making more extract. On August 19, Banting figured out a new way to deplete the pancreas of the external secretion by stimulating it with the hormone secretin, which caused the pancreas to release its enzymes until it was empty. They then ground up the remaining pancreas, with its islets of Langerhans, to make their extract. They no longer had to wait for the pancreas to atrophy.

Twelve hours after Banting and Best gave this new extract to a diabetic collie so weak from its diabetes it could barely stand, the dog was prancing around the lab. They'd done the impossible: created an extract that reduced blood sugar! It didn't have a name, and it would be another 40 years before anyone identified the chemical structure, but Banting and Best had seen firsthand the life-giving benefits of insulin, the hormone secreted by the islets of Langerhans, the chemical every living animal needs to unlock cells and allow in sugar, the energy of life.

We now know that insulin is a protein that acts as a hormone with multiple functions. It increases the rates of DNA replication and protein synthesis in cells by promoting amino acid uptake from the bloodstream into muscles and other tissues. It also forces arterial wall muscles to relax, thereby increasing blood flow. But its best-known function is to regulate sugar metabolism. It does this by promoting the intake of glucose in fat and muscle cells, which make up about two-thirds of the body's cells. In fat cells, insulin converts glucose into triglycerides for storage. In muscles (and the liver), it converts glucose into glycogen. Type one diabetics don't have this storage benefit, so their body runs out of fuel. In contrast, about four hours after a meal, nondiabetics' glycogen begins to break down into glucose in the liver, providing fuel to the body. The body can also break triglycerides down into glucose to use as fuel. Normally there is remarkably little sugar (glucose) in a person's blood stream. A typical grown man has five grams of glucose—about the weight of a restaurant sugar pack—in his whole bloodstream, which carries around five liters of blood (this corresponds to a blood glucose level of 100 mg/dl when fasting). Since diabetics have no insulin, and their livers cannot store glucose as glycogen, the bloodstream is consequently flooded with sugar after every meal. The excess is filtered out by the kidneys and expelled in the urine.

Banting and Best named their extract "isletin." Unfortunately, the extract soon ran out and the dog died nine days later. As Banting wrote, "I have seen patients die and I have never shed a tear. But when that dog died I wanted to be alone, for the tears would fall despite anything I could do."

In early September, Banting returned to London to sell his house and settle his affairs. He and Edith were not getting along and the dreams of marriage had faded, so his focus was now 100 percent on Toronto. Macleod returned from Scotland on September 21. Like cats anxious to present to their master a mouse they had caught, Banting and Best excitedly showed him the extract they had produced. But the senior scientist, aware of the false hope that science often engenders, insisted they repeat their experiments to prove that the extract, and not something else, was responsible for the reduced blood sugar. They did, and succeeded a second time, after which Macleod agreed that they had found something; however, it wasn't stable, nor was it consistently effective. If it were ever going to be produced as a life-saving drug, it had to be purified and stabilized.

Banting and Best returned to the laboratory. Their work progressed slowly through the fall until Macleod suggested they show that regular injections of the extract could keep a dog alive over time—a longevity experiment. But getting enough insulin to provide a long-term treatment was a problem. What little extract they had was maddeningly hard to come by, not to mention the toll it took on the dogs. Searching for new ideas, Banting had begun reading about the work of earlier diabetes researchers and this led to the key step in the whole discovery. He learned that the pancreases of newborn and fetal animals contained more islet cells than other pancreatic cells, since the animals didn't need pancreatic enzymes for digestion until they were born. He also knew from his time on the farm that farmers sometimes bred cattle just before they were killed, to induce them to eat more, get fatter, and fetch a better price. Thus, slaughterhouses often had cattle embryos available.

Banting and Best drove to the cattle yards in northwest Toronto and secured permission to cut the pancreases from nine dead fetuses. From these they produced a new extract. On November 17, they injected dog 27, which had a blood sugar of 300 mg/dl. Within twenty-four hours and after several more injections, the dog's blood sugar had plummeted. This new method of extracting isletin meant Banting's original hypothesis—that part of the pancreas produces enzymes that destroy whatever the islets produce—was invalid. This is common in science, and part of what makes research both so exciting and so maddening; the twists and turns of experiments prove some things and disprove others. But Banting didn't mind discarding his original idea; what mattered was that he and Best had found a better supply of isletin, one that didn't kill tail-wagging dogs. It turns out that insulin in all animals is nearly identical and so can be used across species.

The new extraction method marked the beginning of a new stage of their work. Dog 27 stayed alive until early December, when a too-potent form of the extract threw it into what we now know was hypoglycemic shock (excessively low blood sugar) and it died. Undaunted, the researchers chose another dog, number 33, for their next longevity patient. They also began using alcohol to pull the residue for the extract out of the pancreases. Banting had been pestering MacLeod to add an expert to their team. He had in mind a young biochemist named Bert Collip. Finally, MacLeod could tell by their progress that they needed more resources and Collip joined the team in mid-December 1921. A year younger than Banting, and brilliant at scientific research, Collip brought to the team the technical knowledge of a chemist. He developed a method to test the extracts on rabbits in order to evaluate purity and safety, one that would be used for years. Collip also designed an experiment to show that the liver of a diabetic dog receiving the injections of isletin began retaining the stored form of glucose called glycogen, a key sign of normal glucose metabolism.

Meanwhile, Banting and Best pursued their own experiments. Without telling anyone, they injected each other with isletin. They waited. There was no effect, proving that the extract was not toxic. They also asked Banting's old medical school friend, Joe Gilchrist, who had developed diabetes soon after graduating, to swallow some of the extract. Gilchrist did, but there was no beneficial effect. In fact, developing oral insulin would prove a problem for more than eighty years.

By Christmas, dog 33, now named Marjorie, was still alive, still receiving regular isletin injections, and proving that the extract worked over the long term. Macleod had seen enough. He turned the entire lab over to work on the extract, rechristening it "insulin" from the Latin word *insula*, meaning "island," and brought in Clark Noble, the graduate student who was supposed to work with Banting that fateful summer, but had never started. Best had been so caught up in the work that he had refused his vacation and kept working. Noble focused on the rabbit work and glycogen experiments. Best and another doctor developed tests to determine if the body was actually breaking down and using carbohydrates (a sign that the extract was working), and Collip continued his work on purifying the extract. Just as important, with his original hypothesis abandoned, Banting

277

found that using adult cow pancreases worked just as well as fetal ones. They now had a limitless supply of their most important raw material.

Then came January 11, 1922, and the first human insulin trial on 14 year-old Leonard Thompson. After injecting the brown substance into the boy, the house doctor, Edward Jeffrey, waited an hour, then took a blood sample and rushed it next door to the lab. Within an hour, the boy's blood sugar had dropped from 440mg/dl to 320mg/dl (normal is 100 when fasting). The small improvement was short-lived, however, and an abscess formed at the site of injection, as a result of impurities in the extract. The first human clinical trial of this extract painstakingly created from a cow's pancreas had failed.

But no one was ready to give up. Collip returned to his lab and worked tirelessly to find a way to improve the purity of the extract of bovine pancreas. On the evening of January 16, one of his methods finally yielded results: The proper concentration of alcohol mixed with the extract precipitated out the active ingredient.

On January 23, Leonard Thompson's treatment resumed with the purified extract. Within a day, his urine was nearly glucose-free. His blood sugar dropped from a high of 520mg/dl to 120. Within a few days, the researchers wrote, "The boy became brighter, more active, looked better, and said he felt stronger." Gayle M. Herrington wrote in a 1995 article, "This was the first time in medical history that a diabetic child had been restored to health." Most amazing was that it occurred just eight months after Banting and Best started their work. Thompson lived thirteen more years, finally dying of pneumonia complicated by his diabetes. His pancreas is preserved at the University of Toronto. By February, six more patients had been treated, all with similar results.

But as the work to isolate the active ingredient in the extract and purify it continued that spring, Banting found himself on the periphery of the action. First, Macleod had renamed his discovery, then Collip refused to divulge to him his method of purifying the extract, and finally the Toronto hospital refused to put him on its medical staff, so he was barred from treating patients with the very elixir he had discovered! He began to feel his work was being appropriated, and when he and Edith broke up he started drinking heavily, sometimes stealing the 95 percent proof alcohol used to extract insulin from the pancreases. Throughout March, he later wrote, he never once went to bed sober. He stopped going to the lab, and he spent his days sleeping and his nights smoking his pipe. The other boarders in the house where he rented a small room could often hear him singing "It's a Long, Long Way to Tipperary" and other old war songs.

In May, Macleod decided their research was ready to be presented to the scientific community. They wrote up their findings and were provided a slot at the Association of American Physicians meeting in Washington, D.C., to present their paper. Banting resented Macleod's having taken over the research, so he refused to attend. Only Macleod and Collip traveled south. Called "The Effect Produced on Diabetes by Extracts of Pancreas," the talk marked the first time the word insulin was used in public. It was also the first time the doctors in the audience, including the world's leading diabetologists of the time—Elliott Joslin and Frederick Allen—had ever heard of someone being brought back from the brink of death to actual health. The doctors were stunned. When Macleod finished speaking, they rose as one to give the scientists a standing vote of appreciation, the first ovation of its kind ever at a society meeting.

But Macleod included a word of warning in the paper, alluding to "serious difficulties" the team had found when trying to produce large quantities of the extract. It would prove prophetic.

The Insulin Famine

By the time Macleod and Collip gave their talk, Toronto was in the midst of an insulin famine. It seemed Collip had "lost" the secret for making the extract (he possibly never wrote it down in the first place), and nothing the group did to create a new supply worked. The crisis prompted Best to pay a late-night call to Banting, imploring him to get off the booze. Banting returned to the lab.

The few patients whose lives had been changed by the elixir now began fading. One little girl, the one who was the sickest, drifted into a coma as working supplies ran out. The low potency insulin the team was able to prepare brought her back temporarily, but once it ran out, she died. To make matters worse, news of Macleod and Collip's presentation had spread like wildfire through the medical community. From all over North America, parents flocked to Toronto, bringing their sick and dying children. Diabetics "swarm from all over," Banting wrote to Best in

July, when Best was out of town. "They think we can conjure the extract from the ground." Those with money offered to pay any price to cure their child. Banting even turned down one man's offer of one million dollars for the patent. No amount of money could conjure up enough insulin.

Finally, Best managed to produce the first batch of potent insulin the group had seen in two months by adjusting the acidity of the mixture and using acetone to extract the insulin instead of high heat. He promptly handed it to Banting, who had since opened a private office in Toronto. Banting used some of it to treat his old medical school friend Joe Gilchrist, this time injecting it directly into his bloodstream.

In August, Frederick Allen wrote Banting from New Jersey after receiving an allotment: "I take pleasure in informing you that our first results with your pancreatic extract have been marvelously good. We have cleared up both sugar and acetone in some of the most hopelessly severe cases of diabetes I have ever seen. No bad results have been encountered either generally or locally. We have been able to increase diets, and already an effect seems evident in the form of increased strength…."

The Toronto group realized they could not produce enough insulin for the world, so it was time to turn to industry. The pharmaceutical company Eli Lilly of Indianapolis had contacted the Toronto group back in December when its research director, George Clowes, heard rumors of the research. At the time, Macleod, "suspicious of a big American drug company," told Clowes the university wasn't interested in contracting out the research to America. Now, however, the Toronto group realized the tremendous need to make a large, dependable supply of the drug. The facilities at Toronto simply weren't up to the task. The university agreed to give Lilly a year of exclusivity to manufacture and sell the extract in the U.S. It also took out a patent for insulin in the name of Best and Collip. Banting, who believed that patenting a discovery would go against the Hippocratic Oath, refused to add his name. Under the oath, he felt, it was incumbent on him to make all medical advances "freely available to mankind." (His name was later added.)

Lilly quickly ramped up production. More than 80 years later, diabetes treatments remain at the core of Lilly's products, contributing significantly to the company's multi-billion dollar yearly sales. That same year, 1922, the Danish Nobel laureate August Krogh came to Toronto to learn the process for making insulin. He returned home and created the Nordisk Insulin Laboratory. Today, it is the global pharmaceutical company of Novo Nordisk. Insulin still makes up more than half of its sales.

A "Truly Miraculous" Drug

Twenty-two-year-old Jimmy Havens received his first injection May 22, 1922 in Toronto, thereby becoming the first U.S. citizen to receive insulin. He became an artist specializing in woodcuts and lived until 1983.

The U.S. Secretary of State's daughter, Elizabeth Hughes, received her first injection of insulin with the thick, coarse needles of the day on August 16, 1922. Fourteen years old at the time, she weighed 44 pounds and had the distinctive rounded belly of starvation. After five weeks, she was eating anything she wanted, taking in 2,500 calories a day, and had gained 10 pounds, with no sign of excess sugar in her urine. As she wrote her mother in September: "I look entirely different, everybody says . . . gaining every hour it seems to me in strength and weight. It is truly miraculous." She lived another fifty-nine years, raised three children and traveled throughout the world before dying of a heart attack likely related to her diabetes.

Insulin was even able to bring diabetic patients out of comas, the first being 15-year-old Elsie Needham, who lived another 22 years.

And Teddy Ryder, a 6-year-old who weighed 26 pounds when he received his first insulin injection in 1922, was "fat and happy and thriving on insulin" a year later, his mother reported. He lived another 71 years and became the first human to survive 70 years on the therapy.

As Steven Hume wrote in his book *Frederick Banting: Healer, Hero, Artist*: "No single event in the history of medicine had changed the lives of so many people, so suddenly." The discovery was tantamount to a stay of execution for these children. Suddenly, thousands of young people could actually be young again. They could eat cake on their birthdays, gorge themselves on fried chicken during picnics, and complain of feeling too full. They could once again run and play, learn and study, and grow—as long as they had one thing: daily injections of insulin.

"By Christmas of 1922," wrote Elliott Joslin, who went on to found the world's foremost diabetes clinic, which still bears his name, "I had witness(ed) so many near resurrections that I realized I was seeing enacted before my very eyes Ezekiel's vision of the Valley of Dry Bones:

> . . . and behold, there were very many upon the valley; and lo, they were very dry. And he said to me, 'Son of man, can these bones live?' . . . And as I looked, there were sinews on them, and flesh had come upon them, and skin had covered them; but there was no breath in them. Then he said to me, 'Prophesy to the breath, prophesy, son of man, and say to the breath, Thus says the Lord God: Come from the four winds, O breath, and breathe upon these slain, that they may live.' So I prophesied as he commanded me and the breath came into them, and they lived, and stood upon their feet, an exceedingly great host."

Biostatistician Amy Pearce estimates that by the 1930s over 50,000 children were diagnosed with diabetes and most received treatment through insulin therapy. By 1945, childhood death from diabetes had become a rare occurrence. Based on U.S. census data and prevalence and incidence rates for type one diabetes, it can be estimated that over 136,000 children's lives were saved in the 1950s, followed by over 300,000 for each decade since, resulting in close to 1.4 million American children spared from death between 1922 and 2000. Of course, worldwide the number is much greater. Insulin therapy has saved well over 16 million lives worldwide to date and will continue to save more for years to come.

A Treatment, Not a Cure

As Banting and his team quickly learned, insulin was not a cure for diabetes, but a treatment for its symptoms. Within a few years of its discovery, as the first childhood diabetics began living into adulthood, doctors started seeing an increasing number of cases of blindness, heart disease, kidney disease, and nerve damage in people with the disease. As Joslin later wrote: "The era of coma as the central problem of diabetes gave way to the era of complications." To this day, insulin therapy requires that type one diabetics monitor their blood sugar daily, usually by pricking their finger for a blood test. Their goal is to maintain their blood sugar levels within a small range, not too high, not too low. Blood sugar that is too high is known as hyperglycemia and can result in great thirst, hunger, urination, and blurred vision. If it is too low, the condition is called hypoglycemia, and can lead to confusion, unconsciousness and coma. Type one diabetics give themselves daily shots of insulin to process the sugar in their food (otherwise their blood sugar will be too high), but they also must beware of having too little sugar in their blood, especially if they exercise too much. If their blood sugar gets too low, they must eat sugar.

Today, about one in 500 children gets "childhood" type one diabetes and must take insulin to live—a total of about 30,000 American children each year. Meanwhile, about one in ten adults will develop type two diabetes. Type two diabetes is a very different disease. People at first still make insulin but their bodies have difficulty using it. Later, they may lose their ability to produce insulin as well. It also carries the risk of serious complications like nerve damage, blindness, stroke, and heart disease. Sometimes a byproduct of obesity, its sheer frequency in most Western countries today makes it a serious problem, one that threatens to overwhelm health-care systems. The World Health Organization estimates that over 150 million people have it worldwide, and while some can manage it with diet and exercise, others need oral medication and, eventually, will need insulin to remain alive. More than 17 million have one of either type in the U.S., and more than 25 percent use insulin.

Research on Insulin Has Never Stopped

In the mid-1950s, the British biologist Frederick Sanger identified insulin as a protein and revealed its chemical structure. In 1967, Dorothy Crowfoot Hodgkin determined the molecule's spatial shape using x-ray diffraction (it is stored in a crystalline form in the islets of Langerhans). The two each received the Nobel Prize for their work. Sanger later collected another Nobel Prize for his work on DNA sequencing, the core technology that enabled scientists to sequence the human genome.

Other researchers have found that the insulin molecule is remarkably conserved from an evolutionary perspective. It is almost completely identical in all animals, from fish to fowl to worms to mammals. Pig insulin and human insulin differ by only one amino acid. Cow insulin differs from ours by three. This close molecular similarity explains why cow insulin worked for Frederick Banting just as well as dog insulin. In the decades following its discovery, insulin was obtained from cows, pigs, and fish for the daily shots of diabetics.

Since it was very difficult to make animal-extracted insulin absolutely pure, people allergic to any of these animals could have an adverse reaction to the impurities. So for decades after Banting's discovery, researchers tried to create synthetic insulin. They finally reached their goal in 1978 when Herbert Boyer's laboratory at the University of California at San Francisco inserted a version of the human insulin gene into bacteria and turned the bacteria into little insulin factories. The following year, Genentech Corporation began producing a synthetic insulin, Humulin, the first genetically engineered drug, using recombinant DNA technology. Humulin launched the new industry of biotechnology. In 1982, the Food and Drug Administration approved Humulin for the market, and today it is the primary form of insulin in use. Since 1982, different forms of insulin—long-, short- and medium-acting—have been introduced, enabling people with diabetes to fine-tune their medication depending on what they are eating and how much they are exercising on a particular day.

A year after the introduction of Humulin, the first insulin pump was introduced. About the size of a pager, it was connected to the body via a catheter that had to be changed every few days. Since then, the technology has improved to the point that the pump is now implanted, delivering insulin as needed and eliminating the need for shots and careful measuring of blood sugar levels. About 20 percent of type one diabetics use a pump. In 2006, Exubera, an inhaled therapy, became the first non-injected insulin to be approved by the FDA.

A cure for diabetes is still being sought. Researchers have had some success with kidney-pancreatic transplants and with implants of islets of Langerhans. But there is a shortage of organ donors, and transplants often cause immune rejection, so some researchers have tried to clone a patient's own islets before they stop producing insulin entirely, with the goal of reinserting them back into the pancreas. Other scientists are trying to use embryonic stem cells (which are capable of differentiating into any type of cell) to grow new islet cells identical to an individual's own.

The Nobel Prize

On October 26, 1923, the Nobel Prize in medicine was awarded to Banting and Macleod, the fastest the Nobel committee has ever awarded a prize after an initial discovery. Upon hearing that he was to share the prize with Macleod, however, Banting became very upset that Best was not included, swearing that he wouldn't accept the award. Banting's friends and colleagues converged, calmed him down, and reminded him that he would be the first Canadian to receive the prize, making it a major honor and achievement for both him and his country. Banting grudgingly agreed with his friends that he should not refuse the prize. One thing he could do, however, was split his share of the $30,000 monetary prize with Best. Fifteen thousand dollars was a lot of money in those days, more than most people made in a year, enough to buy a house and have enough left over for a car. Banting immediately cabled Best, who was presenting an academic paper that night at Harvard Medical School, telling him in the truncated language of the wire medium:

> "I ascribe to Best equal share in the discovery stop Hurt
> that he is not so acknowledged by Nobel trustees stop
> Will share with him."

A few days later, Macleod, not to be outdone, announced he would share his portion of the prize with Collip. However, the exclusion of Charles Best remains, as one historian writes, "one of the worst mistakes made by a Nobel committee."

With adulation came academic envy, and many decried Banting's lack of credentials as a diabetes expert. But in the end, the secret to science isn't in the accreditation, but in the results. Thousands of scientists had studied diabetes for more than 100 years, trying to find the secret to the "sugar disease," yet it took a farm boy who was going broke as a doctor, and working in a broiling room that could barely be called a laboratory on the top floor of a Canadian university building, to finally discover the world's first truly life-saving drug. D. E. Robertson adroitly and succinctly countered the complaints about Banting's qualifications by comparing him to Walter Campbell, the young doctor who handled the diabetic patients at Toronto General Hospital. Robertson said, "Campbell knows all about diabetes but cannot treat it; Banting knows nothing about diabetes but can treat it."

Others argued that the specialists in the lab should have gotten more credit. But perhaps Bert Collip himself put it best when he said in later years: "Credit should be apportioned 80 percent to Banting, 10 percent to Best, and 5 percent each to Collip and J. J. R. Macleod."

Banting was awarded a high administrative position at the University of Toronto, his own research institute, a lifetime research grant from the Canadian government, and knighthood in the British court. He married in 1924 and had a child, William, in 1928. The marriage ended after four years, and he married again in 1937. Alarmed by the rise of Nazi Germany, Banting spent the late 1930s involved in military research, including work that lead to the development of the G-suit for the British Royal Air Force. In March 1941, Banting was to fly to England for military work, a precarious excursion in those days with the war going on and aircraft being a still relatively new technology. He expressed premonitions to several friends, but felt duty-bound to go. The bomber on which Banting was flying crashed in the wilds of Newfoundland. Banting managed to dress the wounds of the pilot, Captain Joseph Mackey, but then began lapsing into and out of consciousness. Mackey snow-shoed out of the wilderness to find help, but by the time the rescuers arrived, Frederick Banting was dead. The purpose of Banting's mission is unknown to this day.

Frederick Banting was not a saint, nor perhaps a traditional scientist, but it is rare for one to make history by following all the rules. Indeed, those very traits that made some doubt Banting's acclaim in science might have been the keys to his success. Scientific discovery is not just a matter of disciplined research, as the many investigators who failed to turn up a cure for diabetes demonstrated perfectly. Science is fueled by creativity and insight, and thus it is perhaps no surprise that Banting, who was an avid amateur painter, and cultivated a relationship with the Canadian painters' society known as the Group of Seven, was the one who unlocked the mystery of insulin. Who is the hero of this story? Just ask Teddy Ryder, who scrawled with his small hand:

Dear Dr. Banting,

I wish you could come to see me. I am a fat boy now and I feel fine. I can climb a tree. Margaret would like to see you.

Lots of love from Teddy Ryder

Lives Saved: Over 16 Million

Discovery: Insulin

Crucial Contributors

Frederick Banting: Initiated the research, identified the unpurified extract and moved the research away from dog pancreases to cow pancreases, providing an unlimited supply of raw material from which to derive insulin.

Charles Best: Contributed from the beginning; developed the final method of purifying the extract that was used for years.

Bert Collip: Conducted the initial work of purifying the extract that allowed the first human tests.

John Macleod: Once the extract showed promise, supervised a rigorous and methodical medical study of the extract that enabled it to become a drug.

Charles Best and Frederick Banting on the roof of a University of Toronto building with one of their dogs that had been given insulin (1921). Photo courtesy of the Thomas Fisher Rare Book Library, University of Toronto.

Twenty-eight days passed when Frederick Banting opened his own medical practice in London, Ontario before he had his first patient. His first month totaled $4 of income and one patient.

When Frederick Banting was researching diabetes he had to sell his car in order to buy more dogs to continue.

The first people to ever receive insulin were Frederick Banting and Charles Best. They injected insulin into each other's arms to see if it was safe on humans.

Banting was offered $1 million dollars and royalties by an American financier, but turned it down and never profited from insulin.

Banting wasn't known as a scholar. In his original first notes he spelled diabetes as *diabetus*.

Many diabetes scholars were envious of Banting's success, but D.E. Robertson succinctly countered the complaints about Banting's qualifications, explaining: "Campbell knows all about diabetes but cannot treat it; Banting knows nothing about diabetes but can treat it."

DEAR DR. BANTING, I WISH YOU COULD COME TO SEE ME. I AM A FAT BOY NOW AND I FEEL FINE. I CAN CLIMB A TREE. MARGARET WOULD LIKE TO SEE YOU. LOTS OF LOVE FROM TEDDY RYDER

Teddy Ryder with his sister, Margaret. Teddy weighed only 27 pounds when he began insulin shots at age five. The picture was taken a year later, in 1923. He lived over 70 years on insulin, not dying until he was 76. Photo and letter courtesy of the Thomas Fisher Rare Book Library, University of Toronto.

Before insulin was discovered, type one diabetes, also known as childhood diabetes, was a death sentence. The only prescription was starvation, which extended life a few years.

"No single event in the history of medicine had changed the lives of so many people, so suddenly."
—Stephen Hume, biographer of Banting, on the discovery of insulin

Famous Celebrities with type one diabetes:
Mary Tyler Moore—actress
Ann Rice—writer
Patti Labelle—singer
Gary Hall, Jr.—Olympic swimming gold medalist
Kelli Kuehne—LPGA golfer
Michell McGann—LPGA golfer
Scott Verplank—PGA golfer
Adam Morrison—NBA player

"I hope his young life will be sunshine, but he will have clouds and storms and mists. Above all, I hope his life will be useful. After all, work is the only thing in life that brings happiness."
—Frederick Banting on what he wished for his son

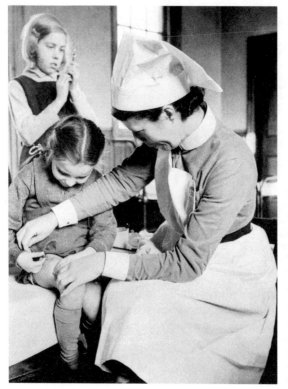

Eileen, a 7-year-old English girl, learning to give herself an insulin shot. The girls behind her are measuring their insulin. Type one diabetics give themselves insulin shots daily. Bettmann Standard RM Collection/Bettman/ CORBIS.

Insulin is absolutely required for all animal life. The mechanism is almost identical in nematode worms, fish and in mammals. In humans, insulin deprivation due to the removal or destruction of the pancreas leads to death in days or at most weeks.

Only about five grams of insulin is circulating in a human's blood, about the amount of a sugar packet.

All insulin came from animals until 1978. Using recombinant DNA, Herbert Boyer at the University of California at San Francisco made a synthetic version of the human insulin gene and inserted it into the bacterium Escheria coli to produce insulin. Boyer's company, Genentech, used this genetically engineered version of insulin to launch the biotechnology industry.

Diabetes used to be called Sugar Sickness. Until 1851, diagnosis was based on the taste of the urine, which may have curbed screening enthusiasm. The physician in charge sometimes called upon the house physician to apply this test, although patients' self-monitoring was preferred.

Chapter 10: Karl Landsteiner— Over 1 Billion and 38 Million Lives Saved

The Superman Scientist: Discoverer of Blood Groups

"As SOON as the blood entered his veins, he felt the heat along his arm and under his armpits. His pulse rose and soon after we observed a plentiful sweat over all his face. His pulse varied extremely at this instant and he complained of great pains in his kidneys, and that he was not well in his stomach, and that he was ready to choke unless given his liberty. He was made to lie down, and fell asleep and slept all night without awakening until morning. When he awakened he made a great glass full of urine, of a colour as black as if it had been mixed with the soot of chimneys."

Those were the words of the French doctor, Jean Denys, who had transfused his patient, Antoine Mauroy, with calf's blood in 1668. Mauroy would later die, as would many who were transfused with human blood before the era of modern medicine. Doctors were perplexed—sometimes transfusions worked, but often the patient would excrete black urine, followed by jaundice, kidney failure, and death. Autopsies would show that these patients' small arteries and capillaries were plugged with red blood cells. Before the twentieth century and the work of a heroic scientist now almost forgotten, the chance of a transfusion's success was no more predictable than a roll of dice.

Incremental Precociousness

In 1891 Austria, 23-year-old Karl Landsteiner looked like an American cowboy in a bronze sculpture by Frederick Remington. He had the stoic expression and downturned mustache, although people in Austria probably thought the facial hair was modeled more on Nietzsche than the wild American West. Remington's paintings of cowboys and Indians also reflected a solitude that characterized Landsteiner well, although we can only speculate as to why, since little is known of his youth. His father died when he was but 6, an event which almost certainly influenced the young Landsteiner's demeanor. Raised by his mother in a house full of books and music, he had no siblings to hear his piano playing or to socialize joyfully with, so perhaps he set his eyes into a hard stare in his photographs out of insecurity or self-consciousness. But there is the chance as well that his solitude stemmed from a secret he held inside—that he was alone in the world of thought, a genius.

Precociousness seemed to grow incrementally as Landsteiner proceeded with his education, almost as if he were consciously taking steps toward brilliance. In gymnasium (high school) he was not at the top of his class, but he was always near the top, moving up year by year from 15th to 8th to as high as 5th in his final year, out of a class of twenty-five. At Vienna University, he took a wide range of courses, and as his future colleague Philip Levine noted, "At 21 years of age he had already known that which had impressed me so much…that important discoveries will emerge when one scientific discipline merges with another."

When he graduated from medical school in 1891, Landsteiner's inquisitive, open mind had hardly begun to be filled. He knew that research, and not clinical practice, was the path he wanted to pursue, so he did what was common at the time (and is somewhat similar to what postdoctoral students do in the United States today): He sought out a famous scientist to study under. But rather than choosing to be the disciple of one famous mentor, Landsteiner traveled during the next three years to study under three of the most famous chemists in Europe, including future Nobel Prize laureate Emil Fischer, who was famous for his work on purines and sugar, and who would later split proteins into amino acids.

If Landsteiner didn't already know he was a genius, he at least practiced Thomas Edison's formula that "genius is 1 percent inspiration and 99 percent perspiration." Wherever he went, his burning desire to learn made him a voracious reader of academic articles and a voluntary partaker of courses that weren't required. And he could not be kept out of the laboratory; as soon as he arrived, he began collaborating with his teachers on experiments that led to published papers.

After being thoroughly immersed in chemistry, Landsteiner studied under a surgeon for a year, and then returned to Vienna University to work under the tutelage of bacteriologist Max von Gruber. Born Jewish, Landsteiner had converted to Catholicism along with his mother when he was in college, well aware that only Catholics could become professors in the Austro-Hungarian Empire (he confirmed he was still Catholic almost fifty years later). Nevertheless, he was passed over for a routine job. Meanwhile, across the campus at the university's Institute of Pathological Anatomy, resided Anton Weichselbaum, a strict and severe bacteriologist famous for discovering the bacterial cause of meningitis. Landsteiner approached him to be an unpaid assistant, was accepted, and immediately started experiments which were to lead to publishable findings. After a year, he applied for a paid job. The Erster Assistent (first or best assistant) at the institute, a powerful person in those days, wanted the same job and told Weichselbaum that he would leave if Landsteiner received the appointment. Weichselbaum went home and thought about it over night, then returned the next day and told him, "You can resign. I am appointing him." Humbled, the Erster Assistant abjured and stayed, and eventually became an admirer of Landsteiner.

Constantin Levaditi, a professor at the Institute Pasteur in Paris, met Landsteiner a little after this time and provides a sensitive description of him: "Now I found myself face to face with a tall, very slim, good-looking young man with brown hair and eyes, a moustache, and a rather sensual mouth. His movements were brisk, yet graceful. We had a brief conversation, in the course of which I was impressed by his extraordinary attraction and by his burning enthusiasm for research-work, despite the unpretentious modesty with which he spoke of it."

In 1899, at the age of 31, Landsteiner received his habilitation in pathological anatomy. The habilitation is a degree above the doctorate that exists in many European countries and is often required to become a full professor. Landsteiner never became a full professor, but he now had paid tasks of teaching university classes, instructing foreign physicians in pathological anatomy, and plenty of time in the dead-house, performing duties such as autopsies, duties which are not likely to make a young man optimistic about life. Every day, he cut into dead bodies and analyzed why they had died. But while pathology was what he was paid for, it was research for which he had prepared himself.

Blood of Our Fathers

Blood has fascinated mankind for centuries, and its mythology permeates every culture and religion. Romans were so convinced that blood carried the vital essence of a person that gladiators drank the blood of their slain opponents. We talk of the "blood of our fathers," "the nation's blood," "blood feuds," and "lifeblood." We become "blood brothers" by smearing together cut thumbs, and Christians drink red wine or grape juice as a stand-in for the blood of Christ during communion.

Hippocrates taught that all living matter contains four "humors"—blood, phlegm, yellow bile, and black bile. Any illness stems from an imbalance of those four elements. Over the next twenty-three centuries, bloodletting—either with the aid of leeches or by simply opening a vein—was often the preferred medical treatment, as it was thought to realign the humors. Bleeding was also thought to be a way to rid the body of its "evil" humors, since blood was thought to be the site of madness. This barbaric approach was not fully abandoned until the 1920s, when the composition and function of blood was finally deeply analyzed, and it was demonstrated through successful transfusions that adding blood could save a lot more lives than removing it.

The science of blood took its first tentative step in 1613, when William Harvey demonstrated that blood circulates in the body, with the heart as its pump (the Arab physician, Ibn al-Nafis, had actually discovered this four hundred years earlier, but his writings never spread). Harvey's discovery overturned the long-accepted model of ancient Greeks such as Aristotle and Galen, who thought that venous blood was born in the liver and bright red arterial blood originated in the heart. From there it supposedly washed out to all parts of the body to be consumed, and some believed that its movements were like the tides of the sea, washing forward and backward.

In the early 1800s, James Blundell, a British physician and obstetrician became, in his words, "appalled at my own helplessness at combating fatal hemorrhage during delivery." Indeed, hemorrhages were responsible for about 22 percent of maternal childbirth fatalities at the time. A compassionate man, Blundell was unhappy with much of the medical practice of his day, which consisted largely of watching people die. He sought out new techniques, including performing the first hysterectomy and pioneering the practice of tying off the fallopian tubes after a cesarean section to prevent a woman from having any more children. He became famous when he proposed transfusing blood from person to person. At the time, the only transfusions had been of animal blood, which he argued against, mostly because it was so impractical: "What is to be done in an emergency? A dog might come when you whistled, but the animal is small; a calf might have appeared fitter for the purpose, but then it had not been taught to walk properly up the stairs!"

Blundell performed the first successful human-to-human blood transfusion in 1829. A woman who was dying from childbirth hemorrhage was transfused with the blood of Blundell's assistant over three hours, and Blundell was happy to report that "the patient expresses herself very strongly on the benefits resulting from the injection of the blood; her observations are equivalent to this—that she felt as if life were infused into her body."

Blundell encouraged other obstetricians to perform transfusions, but the procedure never caught on. In 1849, C. H. F. Routh reviewed all the published transfusions to date and reported that he was only able to identify forty-eight cases, of which eighteen had proven fatal, which was "rather less than that of hernia, or about the same as the average amputation." He concluded that air caught up in the blood was the cause of transfusion reactions. Sporadic attempts were also made to transfuse animal blood in Europe, and milk in the United States, but transfusion was not widely pursued because the transfusion reaction was so terrible to watch that most doctors, aware of the paraphrase of the Hippocrates quote "First, do no harm," refused to even experiment with the practice.

Avoiding the Line of Least Resistance

Germ theory dramatically changed medicine right at the time Landsteiner was born. Prior to Louis Pasteur's development of the theory, most hypotheses presented throughout the centuries about disease could at best be called wild guesses. They were not anchored in a foundation of basic knowledge of how the human body or the natural world works, and moreover they nearly always contained large explanatory gaps between causes and effects. Such common beliefs as spontaneous generation (that disease-causing organisms form from decaying matter), miasma (that disease rides on bad air), or humoral theory (that disease results from an imbalance of bodily humors) withered away as germ theory allowed scientists to identify one specific microbial agent as the cause of each infectious disease. Scientists learned to observe a germ's effect, such as the release of toxins; determine the means by which the toxins cause illness; and thus piece together a causal chain ending in death.

A measure of the importance of a new scientific theory is to observe how many new doors it opens to wide-eyed scientists. Germ theory was like opening the door to a brand new house. A few steps inside, past a nook, was the bathroom where Joseph Lister promulgated antiseptics. If germs are microorganisms, could they not be killed by chemicals or avoided by cleanliness? On a shelf sat the mouthwash, Listerene, named after Lister. Inside the kitchen, Louis Pasteur himself was puttering around, reasoning that food could be decontaminated by heating it and killing the germs residing in it (pasteurization).

Landsteiner began his career amid a continual stream of such new medical insights, and, just as he was entering college, Louis Pasteur opened the Pasteur Institute in Paris that offered a course entitled "Cours de Microbie Technique" ("Course of Microbe Research Techniques"), the first-ever course in microbiology. Microbiology was the most exciting room in the house. In it, Robert Koch discovered the germs that cause tuberculosis and cholera. He could wave you in and show them to you under his microscope. And if you could see a bacterium grow, you could see it die, which offered great hope for a cure. Just as Landsteiner was obtaining his medical degree, Robert Koch published "Koch's postulates," four criteria that are used to establish a causal relationship between a specific microbe and a disease. As countries throughout Europe turned their universities into medical research centers, it is easy to understand how Landsteiner got swept up in the excitement of the medical revolution.

But Landsteiner did not seek to become a microbiologist. Philip Levine wrote of Landsteiner: "The line of least resistance was to embark on efforts to discover new bacteria, now that the methods of cultivation of bacteria were laid down by the discoveries of Pasteur and

Koch. Those who took the line of least resistance did discover the pneumococcus, meningococcus..., but what else did they contribute to the medical sciences?" Landsteiner maintained his belief that chemistry was integral to medicine, and he became intrigued by something more mysterious than germs.

By the 1890s it was known that germs can kill humans by producing toxins inside the body, and that sometimes these toxins were destroyed in the blood, rendering them harmless. Looking into a microscope, scientists could also see bacteria themselves die in blood. What was in blood that could destroy toxins and kill bacteria? These substances would eventually be named "antibodies," and their discovery opened the door to an entirely new room in the house: immunology. What were these agents in blood, so tiny that they could not be seen, so powerful that they worked even when greatly diluted, so specific that they could kill one and only one species in a batch of bacteria? This was the room in which Landsteiner chose to spend his research life.

I Selected the Simplest Experimental Arrangements

The Institute of Pathological Anatomy must have been an inspiring place for Landsteiner to work. Less than 20-years-old, the building was more than three stories tall, made out of monument stone, and looked much like a Washington, D.C. museum. Dedicated to the "location and causes of disease," the institute housed research in a wide range of medical disciplines. Inside, in his white lab coat, his autopsies completed for the day, Landsteiner would sit at a wooden table, the surface shiny, the sunlight reflecting off its varnish. Wooden six-pack racks of test tubes were lined up on the table and a microscope was at hand. Across the room were a centrifuge and containers of dyes. Around Landsteiner were other scientists, some well-known, and myriad assistants, all busy adding to the body of knowledge of the then only decades-old science of medicine.

Early on, microbiologists saw that bacteria sometimes died and dissolved when their cell membranes were broken, a phenomenon called lysing. In 1896, the bacteriologist under whom Landsteiner was studying, Max von Gruber, demonstrated that mixing cholera bacteria with a blood serum from a patient who had developed immunity from cholera resulted in the clumping of bacteria. Clumping had been observed before, but Gruber's experiment was one of the first demonstrations of an immune response to bacteria in blood. Two years later, another scientist mixed different animals' blood and saw that one animal's red blood cells would clump when mixed with another animal's serum. The ability of blood serum to clump bacteria as well as blood cells brought the phenomenon to the forefront of European science. Agglutination, named from the Latin *agglutinare*—"to glue to"—became of great interest, especially since, with blood cells, it could be viewed in a test tube with the naked eye.

Landsteiner was intrigued by agglutination as well. In 1900 he published a paper describing a multifaceted experiment that tested the ability of different animals' blood sera, including that of humans', to inhibit enzymes. In one part of the experiment, Landsteiner observed agglutination of human red blood cells when mixed with both animal and other human blood serum. From the experiment and from reading others' work, he would later write that he learned of an "important general discovery in protein chemistry, namely, that proteins in various animals and plants are different and are specific for each species…. The discovery of biochemical species specificity prompted the question as to whether individuals within a species show similar, though presumably slighter, differences."

It was this discovery along with his observation of agglutination that prompted Landsteiner to add a foreshadowing footnote to the paper: "The sera of healthy individuals not only have an agglutinating effect on animal red cells, but also on human red cells from different individuals. It remains to be decided whether this phenomenon is due to individual differences or to the influence of injuries or bacterial infection."

Thinking on this into 1901, Landsteiner designed an experiment, reasoning that "since no observations whatever had been made in this direction, I selected the simplest experimental arrangements available and the material which offered the best prospects. Accordingly, my experiment consisted of causing the blood serum and erythrocytes [red blood cells] of different human subjects to react with one another."

In hindsight, this might seem an obvious step, but at the time all human blood was thought to be uniform. If it were different, it would also upset the paradigm that was forming about the newly discovered immune system. As a 2003 *British Journal of Haematology* article explained, "At the end of the nineteenth century, cellular and humoral immunity had only one purpose, to defend the body against foreign invaders such as bacteria. No use other than to provide the basis of survival was seen." If the immune system had a more general function than combatting disease, it was far more complex and confusing than had been thought, almost as if it were a funhouse room with concave, funny mirrors.

You can imagine Landsteiner's lab associates, on hearing of another blood experiment, trying to arrive late in the morning or slipping off early to lunch—anything to avoid another stalking scientist wanting to siphon out their blood. Landsteiner eventually gathered blood from five associates, most of them doctors in the lab, as well as himself. Then he set up the straightforward experiment that could be run in any high school laboratory, except that today it would require a phlebotomist to draw the blood and the students would have to wear hazmat suits to protect them from potential HIV or hepatitis.

Each vial of human blood Landsteiner held was 55 percent serum and 45 percent cells, of which red blood cells were the most numerous. Red cells, which function for about 120 days before being replaced, have the vital role of delivering oxygen to other cells throughout the body. The liquid component, serum, is 90 percent water, and was already known to carry the agents that fight disease, which we now refer to as antibodies. Landsteiner first centrifuged each individual's blood at a soft-spin rate, which separated the blood into three layers, with the denser red blood cells at the bottom of the test tube, the white cells and platelets above them in what is today called the "buffy coat layer," and the serum at the top. He decanted the yellowish and clear serum layer into another test tube, then isolated the red blood cells and washed them in a sterile saline solution that removed about 99 percent of the serum proteins and antibodies. After adding a saline solution to dilute them to about 5 percent concentration, he let them sit.

Landsteiner affixed labels to the test tubes containing each individual's serum and red blood cells; he put the sera into one wooden six-pack test-tube rack and the red blood cell test tubes into another. Into a group of six empty test tubes he placed a .6 percent saline solution.

Then, with a pipette, he drew up a little serum of the first person and added it to a test tube with the saline solution. He repeated the procedure for each individual, and then added a few drops of his own red blood cells to all six test tubes. He waited. Nothing happened. He waited longer. The process of agglutination can be seen with the naked eye, but there was no agglutination to be seen.

He set up another six-pack of test tubes with saline solutions, pipetting each individual's serum into each solution. Then he dropped Dr. Sturli's red blood cells into each test tube. The first test tube showed no reaction (it was a mixture of Sturli's own serum and blood cells, so this was no surprise). But after a few minutes in the next test tube, the blood cells began clumping together—agglutination was occurring. Landsteiner's eyes sparkled, but he did not draw any conclusions. To confirm the agglutination he pulled over his microscope and performed a "hanging drop" test. Fastidiously, he placed a small drop of mineral oil on each corner of a cover slip. In the center, he put a drop of the agglutinated blood. Next, he took a microscope cavity slide, inverted it, and dropped it onto the cover slip, gently pressing them together so that the mineral oil spread out and formed a complete seal. Turning the slide over so that the cover slip was on top, he was ready to inspect the drop, which was now hanging freely.

What Landsteiner saw is best described in the words of a scientist from his era: "Under the microscope the cells are close together and cannot be separated by pressure on the cover-slip, but they spread out as if connected by threads. If the pressure is released, the cells move together again." It was agglutination.

He added Dr. Sturli's blood to each of the other sera. Three of the next four test tubes showed agglutination. Landsteiner continued combining all possible combinations of the six individuals, for a total of 36 combinations. He laid out the results in a table, using a dash to indicate no reaction and a plus sign to indicate agglutination. There were sixteen cases of agglutination, six trivial cases of no reaction when an individual's own blood cells were combined with his own serum, and fourteen cases of no agglutination between different individuals.

Table 1. Concerning The Blood Of Six Apparently Healthy Men

| Blood corpuscles of: | | | | | | |
Sera	Dr. St.	Dr. Plecn.	Dr. Sturl	Dr. Erdh.	Zar.	Landst.
Dr. St.	-	+	+	+	+	-
Dr. Plecn.	-	-	+	+	-	-
Dr. Sturl.	-	+	-	-	+	-
Dr. Erdh.	-	+	-	-	+	-
Zar.	-	-	+	+	-	-
Landst.	-	+	+	+	+	-

He called over his associates and Weichselbaum, the head of the lab, showed them his table, and offered them a view of the agglutination in his microscope. The associates nodded their heads at another Landsteiner experiment, gave him a patronizing smile, and drifted back to their research. Weichselbaum gave a jovial "Ja! Ja!" but nothing more.

Landsteiner amusedly remembered in later years that the whole time he was at the lab Weichselbaum never once uttered an "Ausgezeichnet!" (outstanding) upon being presented with a new Landsteiner finding. Only the "Ja! Ja!"

But Landsteiner didn't need acclamation and returned to his work as enthusiastic as ever. He had definitely found something new and significant. Now he would give it the Landsteiner treatment.

Make the Data Thick

Landsteiner had shown agglutination between different humans' blood to be common, occurring almost 50 percent of the time. Did that mean it was random? Since one hallmark of science is cause and effect, it would likely only be random if Landsteiner were making a laboratory mistake. And as Peyton Rous, a later associate of Landsteiner and Nobel laureate, remembered, laboratory exactitude was Landsteiner's signature trait: "Whenever test mixtures were to be made…he would add the crucial component to the tubes and read off the reactions, the assistant standing by. This he did…in greater part to assure himself that what took place actually happened. Of this he was incorrigibly doubtful and when a discovery declared itself he would instantly conclude that it could not be real and would set about to make this plain. In thus striving to pull down, he not only buttressed, but often built further. Experiments which revealed anything were done many times over and not until the data on a point under determination were, in his term, 'thick,' would he publish."

Looking around the lab with needle at hand, Landsteiner knew that to keep on friendly terms with the staff, he should delay sticking any more associates, so he turned to the university hospital and obtained blood from six new mothers. He crossmatched their blood and serum just as he had that of his associates. The results were similar. Seventeen of the thirty-six combinations produced agglutination. Then he crossmatched the sera of the mothers with the red blood cells of the doctors, and obtained a similar result.

Landsteiner continued obtaining blood, and eventually performed a total of 144 crossmatches on blood from twenty-two healthy donors. Finally, he concluded that "the experiments demonstrate that my data require no correction. All twenty-two examined sera from healthy persons gave the reaction." Now he was certain that the observations were true and the phenomenon of agglutination between blood of healthy humans was real.

Well Read

Today Landsteiner would be called a "data hog." With his data set confidently in hand, he proceeded to the second step of the Landsteiner treatment: making certain he was aware of anything and everything known about agglutination. Rous relates that Landsteiner "covered wide reaches in the scientific literature, scanned abstracts of articles for mental relaxation, and was ever the first at the current journals, feeling crestfallen if someone mentioned a pertinent paper which he had not seen. To open mail and break the wrappings of scientific journals gave him eager pleasure." There are references in his papers to obscure academic articles that virtually no one else cited. With such a complete awareness of the current state of knowledge, he well knew the hypotheses that others had proposed as explanations for the agglutination of human blood.

These had to be addressed. One hypothesis was that agglutination was related to the well-known phenomenon of blood clotting. Clotting and agglutination are in some respects similar, so Landsteiner obtained blood from two hemophiliacs, whose blood does not clot normally, and used their sera. Their sera agglutinated with blood of some individuals but not others, just like with healthy donors, so that ruled out a clotting factor.

Then Landsteiner turned to the prevalent hypothesis. Scientists knew that blood that had been exposed to a particular disease thereafter carried something that fought that disease. Agglutination could be a reaction to that substance. But how could anyone ever find human blood that had never been exposed to disease? The answer was deceptively easy, and available nearby. "To exclude the assumption that perhaps past disease processes are of importance, I regarded investigations on the blood of children and animals utilizable." Recently born babies would never have been sick, and you can imagine Landsteiner's enthusiasm as he walked into the lab with fetal placental blood. He prepared the test tubes and mixed the blood cells of the mothers with the sera of their babies—and met with a setback. No agglutination occurred, ever. Landsteiner was forced back to the library. Going through the literature, he found that another scientist had also failed to produce agglutination with serum from infants, and thus Landsteiner concluded that baby blood serum was not developed enough to cause agglutination. Interestingly, and of course completely unknown to Landsteiner, we are not born with these blood-related antibodies in our serum. Instead, they're formed when we're about three months old. Landsteiner next mixed the babies' red blood cells with the mothers' sera. This produced results comparable to those in his other experiments: Out of thirty crossmatches, the babies' red blood cells agglutinated ten times.

Still another hypothesis existed. Several scientists speculated that agglutination was a natural autoimmune response to the re-absorption of constituents of red blood cells. Landsteiner reasoned that if it was an autoimmune response to something red blood cells were releasing, then all the sera should react with what was being released: "Moreover, my investigations show that the different sera do not act identically with respect to agglutination. If one believes, therefore, that they owe their agglutination ability to a kind of autoimmunization through resorption of cell constituents, then one must again assume individual differences to obtain the different sera." Landsteiner did not claim this completely ruled out an autoimmune response, but thought it unlikely to be the cause.

Advancing by One
Limited Hypothesis at a Time

Having ruled out the hypotheses offered by other scientists, Landsteiner only now sought to propose his own explanation. And here, the third component of the Landsteiner treatment served him well. Many of Landsteiner's associates were most surprised that such a brilliant man rarely formed speculative hypotheses.

Dr. Paul Speiser, his Austrian biographer, wrote: "He formulated precisely the relationships between cause and effect, and did not complicate what was still unknown by hypotheses. When he did introduce hypotheses, they were supported scientifically by experiments. He never claimed more than he was able to prove scientifically or could verify by his own experiments."

It is important to recognize that, conceptually, Landsteiner was swimming in a dark pool of blood. What he saw was a completely new blood phenomenon that contradicted the then-current paradigm of the immune system—that its only purpose was to fight germs. Nor did he have the concepts we use today to picture the phenomenon in our minds. Today, we think that the basic function of the immune system is the production of antibodies, but in Landsteiner's day antibodies were such a new and ill-defined concept that they didn't even have a name. In fact, the term *antibody* was only first used that same year in another scientist's paper. The word *antigen*, the other fundamental component of the immune system, did not come into use until 1908.

Landsteiner seemed to have found something new. Some people's blood reacted with others' blood, and some did not. Could it actually be a normal physiological phenomenon? His study of science had taught Landsteiner that if a phenomenon is real, it should offer predictions that can be tested. So he searched through the 144 combinations from the twenty-two donors, looking for a predictable pattern. It was a kind of mathematical problem, and here his genius was fruitful. Rous noted that, "Landsteiner had a mind that was by nature sharp-edged and rigorous, delighting in the exact. He read the higher mathematics for diversion, amused himself with problems in advanced algebra and calculus, and followed with zest each forward step in the new mathematical physics."

The pattern of agglutination reactions Landsteiner had observed didn't appear to be caused by one variable. But what about two variables? Landsteiner noted that "a remarkable regularity appeared in the behavior of the twenty-two blood specimens examined. If one excludes the fetal placental blood, which did not produce agglutination…, in most cases the sera could be divided into three groups: In several cases (group A) the serum reacted on the corpuscles of another group (B), but not on those of group A, whereas the A corpuscles are again influenced in the same manner by serum B. In the third group (C) the serum agglutinates the corpuscles of A and B, while the C corpuscles are not affected by sera of A and B…. In ordinary speech, it can be said that in these cases at least two different kinds of agglutinins [antibodies] are present: some in A, others in B, and both together in C."

When he wrote up the experiment, he titled it "Uber Agglutination-sercheinungen Normalen Menschlichen Blutes" ("On Agglutination Phenomena of Normal Human Blood"). He recognized the importance of his discovery with the last sentence of his academic article: "Finally, it must be mentioned that the reported observations allow us to explain the variable results in therapeutic transfusions of human blood."

Back in the early 1600s, after Harvey's demonstration of the circulation of blood, the German doctor Andreas Libavius wrote: "Let there be a robust youth, healthy and full of vigorous blood; let there stand by him one exhausted of strength, thin, lean and scarce drawing breath; let the master of the art have silver tubes fitting into one another; let him open an artery of the robust person, insert one tube and secure it; let him immediately open an artery of the sick man and insert the other tube; then let him fit the two tubes together and let the blood of the healthy person leap, hot and vigorous, into the sick man and bring the fountain of life and drive away all weakness."

After almost 300 years, on November 14, 1901, it was done. Karl Landsteiner had made one of the fundamental medical discoveries of all time. Human blood, which delivers nutrients, oxygen, and disease-fighting capacity to the human body, is not all identical; it varies among individuals in a fundamental way that allows it to be categorized into a small number of groups. Only specific groups can be transfused. This discovery would make manifest humankind's highest trait—sharing— the sharing of humankind's most valuable commodity, blood.

Today's Blood Group System

A common image used to visualize red blood cell surface antigens is that they are like trees sticking out from the cell surface. Each red blood cell contains millions of antigen trees, which can be either sugars or proteins. If our immune system comes across an antigen it isn't familiar with, it produces an antibody to attack it. The antibody then destroys the cell, sometimes by agglutinating it, often releasing the oxygen of the cell in the process. This is very helpful when bacteria are in our blood, but it is not so beneficial when antibodies go after a transfused blood cell. The agglutination from such an attack is what leads to transfusion reactions, which can occur when our serum's antibodies attack the transfused red blood cell antigens (the most severe reaction) or when the transfused serum has antibodies that attack our own red blood cell antigens (less severe due to the dilution of antibodies in the donor serum).

Landsteiner had found three of the four types of human blood—A, B, and C. Because C contained no antigens, Landsteiner later renamed it 0 (zero), which was later changed to the letter "O" in the U.S., but remains zero in much of the rest of the world. A relentless curiosity pressed Landsteiner to move on to other research. Throughout his life he was intrigued by basic research, leaving the practical applications and refinements of his discoveries to others. Agglutination tests, described as "almost intolerably boring" by some scientists, were indeed too staid for his restless mind. Recognizing that more data needed to be examined to make the discovery "thick," Landsteiner assigned the task to Adriano Sturli who, the next year, working with Alfred Decastello, discovered the fourth blood group, AB. A person with blood type AB has both A and B antigens on the surface of his or her red blood cells. Today, we know that each of us has blood that can be categorized as being part of the ABO blood group. Each of our blood's red cells has either an A antigen on its surface, a B antigen, both, or neither.

Toward the end of his life, Landsteiner, along with former student Alexander Wiener, discovered another important blood group, the Rh (rhesus) group, so named because the researchers were working with red blood cells taken from rhesus macaque monkeys. Wiener and Philip Levine, another of Landsteiner's former students, determined its importance, which explained most of the rare blood transfusion reactions that still occurred. In addition, they saved many babies from the previously mysterious and potentially fatal condition *erythroblastosis fetalis*, or Rh disease, which occurs when an Rh-negative mother carries an Rh-positive baby, and develops anti-Rh antibodies as a result of being exposed to her baby's Rh antigens. During subsequent pregnancies with an Rh-positive baby, the mother's antibodies can react with the baby's red blood cells in a potentially fatal way. Wiener also developed the first treatment for this condition, a so-called "exchange transfusion," which involves removing much of the baby's blood and replacing it with fresh blood. This procedure has since been replaced by the practice of giving the mother an injection of anti-Rh globulin to bind any red blood cells before her immune system can produce antibodies against them.

The Rh group is the most polymorphic of the blood groups known, with at least forty-five distinct antigens, of which the D antigen is the most reactive to the immune system. As a result of its critical importance, all people are now tested to categorize their red blood cells in both the ABO and RhD blood groups. The RhD antigen is noted by adding a plus sign if it is present on a person's red blood cells or a minus sign if it is absent. About 85 percent of the U.S. population has the RhD antigen. Thus all people are classified into one of these eight categories:

U.S. Population

Blood Type	A+	A-	B+	B-	O+	O-	AB+	AB-
% of Population	34	6	9	2	38	7	3	1

Landsteiner's discovery of the ABO blood group opened the door to the discovery of many other blood groups. The ABO and RhD blood groups are only two of at least 26 blood groups now known (Landsteiner also discovered the MN and P groups), and each group is distinguished by antigens on red blood cells that are controlled by a single gene. Most of the blood groups other than ABO and RhD don't cause transfusion reactions because blood serum rarely carries antibodies that will attack these antigens. One of these antigens is the Duffy antigen discussed in the Paul Müller chapter. A majority of people of African descent do not have the Duffy antigen, as it is strongly selected against because it provides a means of entry for a certain type of malaria into red blood cells. However, the function of many of the antigens is still unknown.

Much more is known about antibodies today than in Landsteiner's time. All human blood serum contains antibodies, which are Y-shaped proteins that the immune system creates to identify and destroy foreign objects (sometimes latching onto them and sometimes calling forth other processes to disrupt them). They are produced as a specific reaction to the presence of these antigens, whether on a polio virus, a strep bacterium, or a red blood cell. It is not known what triggers most humans' immune systems to produce A or B antibodies, since our bodies are not always confronted with these antigens prior to a blood transfusion, but all humans have A and B antibodies unless they have the A or B antigen on their red blood cells. Some scientists speculate that there exist bacterial antigens that are similar to A and B antigens. Babies then come into contact with these bacteria and produce antibodies that can also recognize A or B antigens.

The primary risk of blood transfusion reactions with the other twenty-four blood groups occurs after a transfusion or pregnancy, which sometimes causes the body to produce an antibody to the newly acquired antigen, thereafter keeping it in reserve. When a second transfusion is given, or second pregnancy occurs, the antibodies come out to fight. Even so, serious reactions to the other groups are extremely rare.

After 3,639 Autopsies, Marriage

Adriano Sturli, one of the doctors who donated blood for Landsteiner's blood group experiment, told an amusing story about Landsteiner: "Towards the end of 1901, I was asked by Landsteiner, with whom I had up to then only done some histological and bacteriological studies, whether I would like to join him in some serological studies and experiments [these would lead to the discovery of AB, the fourth blood group]. I agreed with alacrity, and thus had an opportunity of repeating all his experiments to his explicit satisfaction. I should add that the final studies started on the afternoon of 31 December 1901 and went on without a break until 8:30 PM. The two of us were quite alone in the Pathological Institute, now silent and deserted. These hours were a sort of tragicomedy for me, because I had naturally been itching for hours to join my friends and see the New Year in style. However, Landsteiner was gently but firmly insistent, and kept me washing blood-corpuscles, mixing sera, centrifuging, saturating charcoal-powder with dyes, etc. under his supervision, with results that seemed to me amazing but to Landsteiner were self-evident. Eventually we took leave of one another, tired but still good friends, with a cordial 'Happy New Year.'"

Indeed, it would be difficult to find another scientist as passionate about research or as thorough in experimentation as Landsteiner. One journalist, tallying up Landsteiner's work hours, concluded that over a fifty-year span he spent 90 percent of his conscious life in scientific pursuits.

In 1908, his work ethic paid off. Landsteiner took over the Department of Pathology at the Imperial Wilhelmina Hospital in Vienna. Finally, after performing 3,639 autopsies, he could leave the dead-house work to an assistant. The Institute of Pathological Anatomy thought so highly of him that it converted a stable into a lab so that he also could continue his research there. Those there remembered him as "a dignified personality with a reserved but kindly manner," and a "shy character who would never initiate a discussion but was always ready to help a colleague and respond to questions or requests." He was known also for his generosity, paying above the minimum for the upkeep of the lab's animals.

Now freed of the autopsies, there were many good days of experiments at the institute. He would bring his beloved dog, Waldi, to work and let him sit under his desk. Around lunchtime every day, Waldi would start barking and Landsteiner would playfully reprimand him, "Waldi, you've not an atom of respect for science." After a few years, he began dating Helene Wlasto and then, as World War I began, they married. Twelve years younger than Karl, she was described as sweet and devoted. They had a son, Ernst Karl, and were joined by her sister to make a lively household.

While almost all accounts of Landsteiner note his social reticence, at work Landsteiner was industrious. Rous, who worked on the floor above him at the Rockefeller Institute, where Landsteiner eventually worked, said that in the lab "he was full of confident energy and enterprise. He had his workrooms equipped as if for chemistry and preferred assistants trained in that branch, not physicians or biologists. Social activities he avoided; in his view, the day was for experiment only, reading and thinking could be done at night, until a late hour.... His energy was continuous and compelling, and no moment of idleness in the laboratory was tolerable to him. To each of his assistants he allotted some phase of one of his problems, to be worked upon separately, and he encouraged the trying out of any thoughts concerning it that the assistant might have. To himself, new ideas came endlessly and he was continually suggesting trial experiments which 'would take no time'. His interest was discovery, not training investigators, but each day at lunch he talked with his group about what they were doing, and when a new fact was brought to light he would call them all together to demonstrate it, asking himself and them what it meant."

One of those students, John Jacobs, contrasted his work in Landsteiner's lab with his previous experience at Harvard University, where he was largely left to his own devices: "In Dr. Landsteiner's department I found myself in a closely knit organization in the German tradition in which all my research plans and technical efforts were closely supervised and instruction was given, in a most kindly way, on every essential point. In such an environment and with such an instructor as Karl Landsteiner it was possible to learn quickly."

Jacobs particularly remembered the process of writing up their scientific discoveries: "Papers were generally written at night on the dining-room table in Dr. Landsteiner's large, very simply furnished apartment. It would begin by presenting a first draft, which, in the course of the evening, would be so criss-crossed with corrections and suggestions that it would have to be entirely rewritten. After three or four such sessions, made more pleasant by snacks prepared by Mrs. Landsteiner and including foreign cheeses and a little wine when work was over, the shape of a paper gradually appeared.... In writing papers Dr. Landsteiner was never ready to put pen to paper until he had definitely established a new fact. The paper was then built about this fact and its relationships discussed... the discussions seemed to me to be unique. This was brought about by the fact that he limited himself severely to pointing out the highly probable implications and relationships of the facts observed, almost completely omitting opinion and theory.... During these long sessions Mrs. Jacobs and Mrs. Landsteiner would often sneak off to a moving picture theatre (of which Dr. Landsteiner disapproved).... Summaries were likewise worded with extreme caution and conservation. A large element of his genius consisted in the humility with which he would forgo the opportunity to draw broad theoretical conclusions in the interest of maintaining a high degree of accuracy and objective reality."

Overall, it was an exhilarating experience for young scientists. Jacob wrote, "We were a closely knit group, united by our interest in these subjects and by our admiration of the 'Chief'.... There was a constant give and take of ideas in his laboratory in which he was the gifted leader, but the accurate observations, the technical execution of experiments and the spotting of new factual findings were largely the responsibility of the assistants, which had great importance and significance, especially in respect to original observation, in Landsteiner's method of investigation. Thus when, after I had been with him something over a year, one day he put his hand on my shoulder and said, 'John, you have a gift for research,' I felt that this was an honor that no degree, title, praise, or honor has or could ever equal."

346 Scientific Articles

If Pasteur opened doors to many rooms with his germ theory, Landsteiner entered them and filled them with trophies. While still researching blood groups, he quickly foresaw the usefulness of his discovery for paternity tests and forensics. He knew from the "hanging drop" test that it took very little blood to cause an agglutination reaction, so he tested a drop of blood he had spilled on a cloth: "The described agglutination can also be produced with serum which has been dried and then dissolved. I did this successfully with a solution from a drop of blood which had been dried on linen and preserved for 14 days....To be sure, on the second test, the six sera in Table I exhibited the same behavior as the specimens taken 9 days earlier." This led to another new conclusion: "Thus the reaction may possibly be suitable for forensic purposes of identification in some cases, or, better, for the detection of the nonidentity of blood specimens."

311

When Landsteiner took up the study of syphilis, he made several key discoveries, the most far-reaching being the application of a microscopic technique known as dark field illumination. Using it, scientists were able to observe syphilis and other bacteria for hours on end, leading to much more knowledge about how they lived and how they could be destroyed. Next, working much of the time with Constantin Levaditi, Landsteiner furnished most of the basic knowledge about polio. He proved that it was caused by one specific agent, and that the agent was a virus, not a bacterium. He introduced a technique for preserving the virus in glycerin, developed a blood test for it, and demonstrated how to transfer it to monkeys. Monkeys would be the primary experimental animal for polio until John Enders' work in the 1940s. Finally, Landsteiner demonstrated that the serum of monkeys who had become ill with polio could inactivate the virus, indicating that a vaccine was potentially possible.

Landsteiner's accomplishments are impossible to list in one chapter. He published 346 scientific articles, many of which significantly advanced the field with which they dealt. Immunology was the field of medical science that most intrigued Landsteiner, and he considered his discoveries in it his most important contribution to science. Besides his blood group research, he also provided some of the first evidence that allergic reactions are in fact immune reactions.

Specificity

Paul Ehrlich was the most famous scientist in Europe in the early 1900s. He was not only a legitimately great scientist, but also media savvy, popular for formulating grand hypotheses that were intuitively easy to understand. For instance, he popularized the notion that science could produce a specific "magic bullet" to fight any given disease. Many people thought Ehrlich had figured out the immune system. He suggested that blood cells have side-chains—hypothetical structures that grow out of the cells when stimulated by toxins. These side-chains are shaped to grab toxins and then lock them away in their grasp. Antibodies found in serum were these structures, broken off from or shed by the blood cells. Just as the solar system model of the atom is hard to dispel to this day, the most popular implication of Ehrlich's theory—that there is a perfect one-to-one antigen-antibody fit that became known as the "lock and key" fit—still persists to some degree today, even though it is mostly incorrect.

Landsteiner, by contrast, was not one to put forth bold, speculative hypotheses. Steadily, year after year, experiment by experiment, he added to the basic body of knowledge about the immune system. In 1909, one year after Ehrlich had received the Nobel Prize, Landsteiner gave a report at the 16th International Medical Congress in Budapest that cited his own and other investigators' research, and concluded that antibody formation is not an "overproduction" of cell structures, as Ehrlich had proposed. Through more than two decades of work, Landsteiner demonstrated that the immune system is much more recondite, labyrinthine, and complex than Ehrlich had thought.

To sum up his immunological discoveries, Landsteiner wrote a book, which was read for decades thereafter by immunologists. He named it *The Specificity of Serological Reactions*, because he was seeking to define how specific antibodies were in attacking pathogens or other substances. In 2001, Herman Eisen of MIT wrote: "Landsteiner's work dispelled any notion that might have once been held that there is absolute specificity in immune relations. Indeed, the structures of the many cross-reacting molecules uncovered in work from his laboratory were used to great advantage to illuminate how a ligand's [any molecule that binds to another] shape, size, and charge distribution affects the extent to which it is recognized by antibodies."

Specificity was so important to Landsteiner that friends often kidded him about it. One wrote to him requesting photographs that would define Landsteiner's "specifistic" nature. Landsteiner returned them, along with an accompanying explanation that the "pictures should define a specifistic, at least if tested against the lower primates."

Let the Blood of the Healthy Person Leap, Hot and Vigorous, into the Sick Man

It is interesting to compare the uptake time of Einstein's theory of special relativity and Landsteiner's theory of blood groups. Both were born in the early 1900's, and both took around fifteen years to become widely accepted.

In the first years after Landsteiner's blood group discovery, not much happened. There existed one major practical hurdle—the fact that blood clotted within minutes after it was withdrawn. To solve this problem, surgeons of the era came up with a seemingly draconian method of transferring blood. They would sew together an artery of a donor and the vein of the patient who needed blood. Even if you could talk someone into being the donor, this was an intricate surgical procedure, made more problematic because the only way to measure how much blood was transfused was to assess the condition of both the recipient and the donor. If they looked better, or worse, than at the beginning of the procedure, the transfusion would be stopped.

In an effort to jumpstart blood group testing, Landsteiner repeated in a scientific publication that transfusion reactions likely stemmed from blood group incompatibility, and eventually Reuben Ottenberg carried out the first blood group compatibility testing in 1907, at Mount Sinai Hospital in New York City. He went on to observe that people with group O blood could be universal donors (an understanding which has since been revised). Up until 1913, only around fifty transfusions a year were carried out in a typical New York City hospital. Then, in 1914, Dr. Richard Lewisohn, also at Mount Sinai Hospital, discovered that adding sodium citrate could prevent blood from clotting. Surgical transfusion procedures could now be abandoned, and blood could instead be drained into a container, and then transfused in a measured amount. The discovery came just in time. The modern warfare of WWI was causing widespread injury and death, much of it from shock and blood loss. With a safe, storable form of blood now available, an American lieutenant serving as a surgeon on the Western Front took it upon himself to build the first blood bank. Oswald Hope Robertson, using some of the ideas of Peyton Rous (who has been quoted in this chapter), collected blood from donors into clean glass bottles containing the anticoagulant solution, and transported them to battlefield hospitals. Tens of thousands of transfusions were performed from 1914 to 1918, and British medical history books declared blood transfusion "the most important medical development of the war."

Thereafter, transfusions became a normal hospital procedure as indicated in a 1921 letter Landsteiner wrote to Sturli: "I was very glad to have your pretty card: thank you very much.... the isoagglutinins we were working on are highly thought of in America, where an immense number of transfusions are performed, the blood-group determined in every case...."

The first civilian blood bank in the world was established in 1932 in Leningrad, Russia. Bernard Fantus, who coined the term "blood bank," established the first one in the U.S. at Cook County Hospital in Chicago in 1937. Cities all over the world quickly built blood banks. After World War II, the Red Cross started a nationwide collection program that now supplies nearly 50 percent of the blood transfused in the United States. In 1950, plastic bags were introduced and quickly replaced glass bottles, making handling blood much easier and safer.

As soon as blood transfusions became common, other risks became apparent. The biggest danger was that the transfusion of blood carried with it a disease from the donor. In Landsteiner's day, before the advent of antibiotics, syphilis was a dreaded contagious disease. In 1906, August Wassermann became famous for developing the first blood test for syphilis. Within a year, Landsteiner demonstrated that the test was not the specific reaction to syphilis antibodies it was purported to be; the test had to be refined. Today, tests screening for bacterial diseases such as syphilis, as well as for viruses such as hepatitis and HIV, have become integral to blood transfusion medicine.

Blood transfusion is now amazingly safe. Fatal blood group reactions occur only about once in 250,000 transfusions. About half of those are due to error, usually incorrect screening or blood given to the wrong person. The risk of getting hepatitis B is about one in 250,000, as well. The risk of getting HIV from a transfusion dropped dramatically in the 1990s, after improved tests were developed and implemented, to about one in two million.

The Greatest Life-Saving Medical Advance in History

According to our research, blood transfusions have turned out to be the greatest life-saving medical advance in history. To tabulate the lives saved by transfusions, Amy Pearce began counting only in 1955, when data began to be documented well. She arrived at the astounding figure of 1 billion, 38 million lives saved. Transfusions are at the heart of critical care, saving lives due to trauma such as occurs in automobile accidents, the victims of which can require up to fifty units of blood. A blood unit is the amount that most people give, which is about one pint. Other major uses of blood are organ transplants (up to thirty units), severe burns (up to twenty units), and heart surgery (up to six units). James Blundell, the nineteenth century obstetrician, would be happy as well. In the developing world where maternal hemorrhaging after childbirth is still common, transfusions are vital to saving new mothers' lives. Blood can also be transformed into at least twenty-five specific blood products used in medicine. Platelets aid in the clotting process, gamma globulins increase immune activity, and plasma (AB plasma is considered the universal blood plasma) is used for trauma patients.

Donating blood is one of the most charitable and beneficial gifts a man or woman can give another person. Almost 4 million Americans receive gifts of blood each year. Americans donate approximately 12 million units of blood yearly, which are processed into 20 million blood products. Worldwide, 80 million units are donated every year. Almost half of the U.S. population has donated blood at some time, making it impossible not to be optimistic about humankind's compassion.

The Melancholy Genius

As World War I proceeded, Landsteiner's career stalled. He was never made a full professor in Vienna, and historians have speculated as to why. Some claim it was due to vestiges of anti-Semitism, while others think it was because he rebelled against Ehrlich. But other scientists with similar problems advanced further, so a more likely culprit was Landsteiner's introverted nature. He could not play the academic social game, either by courting the public to become famous or by socializing with administrators behind the scenes.

When Austria-Hungary began losing the war, living conditions in the empire rapidly deteriorated. Landsteiner bought a goat to provide milk for his son, but a visiting scientist reported that Landsteiner "looked as if he were starving." Electricity became rare in the city, and all the trees around Landsteiner's house were cut for firewood. He tried to soldier on: One of his 1917 experiments involved thirty-three different antigens, twenty-three sera, and 759 test reactions. This enormous project was carried out with no heat in his lab. Experimental animals were malnourished, dooming other experiments. At the end of the war, conditions became even worse, so when Landsteiner arrived home one evening to find the fence around his yard stolen for firewood, he resolved to leave. A professor friend at Leiden University in the Netherlands, Storm van Leeuwen, found Landsteiner a lab tech position at the Catholic Hospital in The Hague in Holland. His neighbor, a teenage boy who had befriended him, remembered his parting words, "Finish your studies and then go out into the great big world. And if you are ever in any trouble let me know, and I'll do my best to help."

So in 1919, at the age of 51, having lost everything, Landsteiner emigrated to Holland, where he did routine lab work, including, once again, dead-house post mortems. He even had to take a part-time job. Working in one small room that he shared with a nun, and that was also the coffee break room, he continued his research unabated, and published twelve papers over the next two years. But van Leeuwen realized that Landsteiner's research was stifled, and so recommended him to the Rockefeller Institute, which invited him to New York. Landsteiner had always wanted his own research institute, and while not promised quite this, for the first time in his life, at the age of 54, he was offered a position that gave him the freedom to devote all of his time to research.

With his family, and a barrel of carcinogenic tar he used to induce tumors in animals, Landsteiner arrived in New York in the spring of 1923 and found an apartment over a butcher shop on Madison Avenue. He delighted at learning new words and as he picked up English quickly, he and his family resolved to stay and become citizens. Nevertheless, there are many accounts of older immigrants having trouble adjusting to a new culture, and Landsteiner was no exception. He had lived in a quiet house by the sea in Holland, and the noise of city life bothered him. When his long-time associate Constantin Levaditi visited, he related, "For many years I had had no news of Landsteiner at all; but during a study-tour of America in 1929, I ran him to earth in his laboratory. I found him rather depressed and full of complaints about his work, especially about the regulations that, apart from restricting his activities in certain fields, bore no relation at all to the virtually unlimited scope of his own scientific ambitions."

It was most likely his inability to research polio that most disturbed Landsteiner. Polio was the pet project of Simon Flexner, who, as the head of the Rockefeller Institute, controlled who worked on it. It is only speculation, but Flexner may have been afraid that Landsteiner would overturn his own firmly held polio paradigm. Since Landsteiner had discovered most of the fundamental knowledge about polio, it is difficult to understand any motive other than envy that Flexner could have had in forbidding Landsteiner to research it. Landsteiner told Levaditi with a wry smile, "If I am asked to make do with only half a microscope I have to comply."

Landsteiner's public persona was always subdued, even evasive. He avoided the media and did not like having his picture taken. He also expressed his difficulty at home. "Just imagine. I can't even play the piano I managed to lay my hands on. It was my only form of relaxation, but the neighbors complained that I made such a noise they couldn't listen to the radio. So now I never touch it." As the Depression encroached into the 1930s, Landsteiner became more and more discouraged about the state of civilization, as well. He was the product of the German scientific tradition, so the death spiral of European history had a strong impact on him, especially as he heard firsthand accounts of the rise of Hitler. Friedrich Schiff, whom Landsteiner called "the foremost German research-worker in the field of blood-groups," was forced to emigrate to the U.S. in 1936 when, ironically, only two years before, Nazi storm troopers had arrived at his laboratory wanting to have their blood-groups identified. Landsteiner inquired about jobs for German professors and also was known to have sent them money.

Not that Landsteiner was always dire. He still relished his lab, never losing his wit while there. The young scientists in his charge often rushed to conclusions. One day he said, "Is it not strange that I, who have so little time left, should be teaching patience to you, who have your life before you?"

In 1939, Landsteiner officially retired, but in a highly unusual arrangement was allowed to keep his lab. Working just as hard, he produced another twenty-eight publications. But time was catching up with him.

Levine relates how one day, "with tears in his eyes," Landsteiner divulged that his wife had cancer and that he hoped to die before she did. He turned to his lab for solace, studying cancer to try to help his wife, but on June 24, 1943, he suffered a heart attack and died two days later. He was 75 and had spent 51 years in the lab. His wife died later that year, on Christmas day, leaving their 26-year-old son, who had become a medical doctor.

Together, all of the events of his last years, as well as his reserved attributes, brought to him the monikers of the "the investigator with the mournful eyes" and the "melancholy genius."

His Was a Sad Pessimism, Never Bitter

It is fitting that Landsteiner's discovery led to the most lives saved in medical history. He was the consummate scientist, contributing primary knowledge to four fields: serology, virology, immunology, and allergy. Michael Heidelberger, a collaborator, said, "The time is past when one man can know all of science. Karl Landsteiner was one of the last possessed of the tremendous intellect that could comprehend and, better still, use practically all of the scientific knowledge of his time."

Landsteiner's research had lasting influence. In the 1970s, scientists developed a flu vaccine known as Hib (for *Haemophilus influenzae type b*). Hib was at the time one of the most lethal pathogens of early childhood in the developed world. The vaccine worked fine for children over two, but it did not work well for infants. Stymied, researchers searched through historical references and discovered a technique developed by none other than Karl Landsteiner. In the 1980s, using Landsteiner's articulation of hapten-carrier theory, Oswald Avery and W. F. Goebel developed the first Hib conjugate vaccine. Introduced into widespread use in 1990, it brought about a 99 percent decline of invasive Hib disease. So every parent whose child gets vaccinated is receiving a small gift from Landsteiner's far-reaching research.

Landsteiner was truly the superman scientist. What's more, in personality he was essentially human: strong in intellect and sensitively insecure in human relations, full of the excitement of discovery and the despair at civilization's frailty, sometimes all at the same time.

One evening in 1930, Philip Levine went to Landsteiner's apartment. He found the family placidly reading as usual. Levine was quite surprised, considering the news of the day. "What news?" Landsteiner's wife and son asked. Their looks told the story. Karl had told neither Helene nor Ernst what he had learned earlier that day—that he had been awarded the Nobel Prize in Medicine. Beginning in 1923, fourteen different nominators had put him up for the award for three different discoveries: his polio research, his immune system work, and his discovery of the blood groups. After 29 years, the committee finally granted him the award for his blood group work.

Landsteiner feigned that he was uncertain it was true. But it was. And true also was what this incident revealed of his personality. Peyton Rous, who would earn his own Nobel Prize 37 years later for discovering tumor-inducing viruses, and who had congratulated him that same day at the lab, later wrote Landsteiner's obituary. He said that Karl "was peremptory by nature but he was downcast, too, self-questioning and never sure in his human relations.... His pessimism was sad, not bitter, and never obtrusive; despite it, he cheered others in what they were trying to do. But he was held in the grip of his temperament and gradually this tightened upon him: Only in his scientific life was he serene and whole. Fortunately, he was sensitive, else he might have been stern and demanding, and he had an eager yet hesitant desire to be liked. Liking came to him, for he was simple, sincere, modest and gentle, and witty as well in a shy way."

Lives Saved: Over 1 Billion and 38 Million

Discovery:　　　　Blood groups

Crucial Contributors

Karl Landsteiner:　　In 1901, he discovered that blood was not uniform amongst humanity. Blood varies such that mixing one person's blood with another's by transfusion can cause a deadly reaction. After the discovery of blood groups and their use became routine for all blood transfusions, reactions dropped precipitously. This allowed blood to be transfused for many medical emergencies, resulting in the saving of more lives than any other medical procedure.

Karl Landsteiner

Richard Lewisohn: In 1913, added citrate to blood to prevent clotting, allowing for the banking of blood for blood transfusions.

Karl
Landsteiner
in his
New York
lab in 1930.
Bettmann
Standard RM
Collection/
Bettman/
CORBIS.

"The time is past when one man can know all of science. Karl Landsteiner was one of the last possessed of the tremendous intellect that could comprehend and, better still, use practically all of the scientific knowledge of his time."
—Professor Michael Heidelberger

A journalist, tallying up Landsteiner's work hours, concluded that over a 50 year span he spent 90 percent of his conscious life on scientific pursuits.

"He was full of confident energy and enterprise. He had his workrooms equipped as if for chemistry and preferred assistants trained in that branch, not physicians or biologists. Social activities he avoided; in his view the day was for experiment only, reading and thinking could be done at night—until a late hour…. His energy was continuous and compelling, and no moment of idleness in the laboratory was tolerable to him…. To himself new ideas came endlessly and he was continually suggesting trial experiments which 'would take no time'."
—Peyton Rous, Nobel Laureate and associate

"Amongst Dr. Landsteiner's laboratory assistants, when speaking to each other, he was affectionately known as the 'Chief'."
—John Jacobs, student

In his Austrian days Landsteiner would bring his beloved dog, Waldi, to work and let him sit under his desk. Around lunchtime every day Waldi would start barking and Landsteiner would playfully reprimand him, "Waldi, you've not an atom of respect for science."

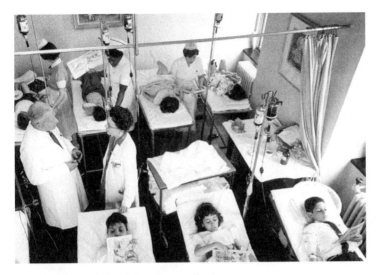

Children receiving blood transfusions.
Bob Gomel/Time & Life Pictures/Getty Images.

"It soon became clear, however, that the reactions follow a pattern, which is valid for the blood of all humans, and that the peculiarities discovered are just as characteristic of the individual as are the serological features peculiar to an animal species...."
—Karl Landsteiner

"Landsteiner's paper of 1901 ranks with the most famous classical writings in all medicine... Landsteiner was one of the greatest scientists of all time. Through his work on blood-groups he became one of the greatest benefactors of mankind. The work of this great son of Austria saved, and continues to save, the lives of millions."
—Neues Ost Juvenal

Blood Transfusion Facts
- At least 25 blood groups have been identified.
- The most dangerous ones were those discovered by Landsteiner and his associates.
- Almost 4 million Americans receive gifts of blood each year.
- Worldwide, 80 million units are donated every year.
- World War I started the wide-spread use of blood transfusions.
- The first civilian blood bank was established in 1932 in Leningrad, Russia.
- Bernard Fantus, who coined the term "blood bank," established the first one in the United States at Cook County Hospital in Chicago in 1937.
- The risk of getting HIV from a transfusion is about one in two million.
- Almost half of the United States population has donated blood at some time, making it impossible not to be optimistic about humankind's compassion.

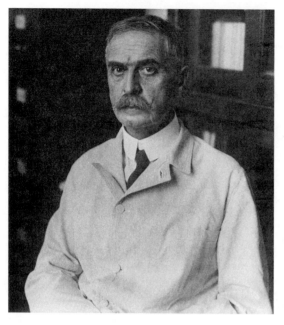

The melancholy
genius. Bettmann
Standard RM Collection/
Bettmann/CORBIS.

When Landsteiner discovered blood groups, his main job was as a pathologist. He performed 3,639 autopsies over the ten years he held that job.

Having seen the devastation of World War I and watching the rise of Hitler, Landsteiner was not an optimist, so his reserved personality brought to him the monikers of the "the investigator with the mournful eyes" and the "melancholy genius."

"The rigid limitation of his experiments to the exploration of facts (avoiding theories)—advancing by one limited hypothesis at a time—kept his work close to objective reality and, no doubt, greatly increased the percentage of positive findings."
—John Jacobs, student and later associate

Landsteiner published 346 scientific articles, and can be considered a founding father in the fields of immunology, polio, and allergy, as well as hematology.

To Landsteiner, the experiment was the excitement. He was an inductive master. The truth was not obtained by guessing, it was obtained by piling up the facts until the facts manifest a new biological function. So like a lone cowboy, he rode off into the unexplored canyons of science and over the next decades made prodigious discoveries.

Chapter 11: A Choice for the Future

Our Health or Our Wealth?

THIS BOOK has been a celebration of the accomplishments of ten scientists. Five of them made their discoveries in the first half of the twentieth century. And what a fifty-year period it was, pushing American life expectancy from 45 years to 65 years. The impact of the other five scientists occurred in the latter half of the century. Four of their discoveries primarily affected the developing world, so it might be concluded that the major breakthroughs have all occurred, and only incremental advances are left. Yet the incremental advances add up in many ways to just as remarkable a revolution, one that should be called the "silent miracle." With few monumental, well-publicized breakthroughs, deaths have continued to decrease dramatically.

Since 1950, life expectancy from birth has increased by 12 years to 77 in the United States. Between 1950 and 1994, even the life expectancy of the average 45-year-old increased by 4.5 years. Think of being middle-aged and someone giving you four-and-a-half more years of life. Furthermore, they will likely be good years: Since 1980, the percentage of the elderly that have difficulty living independently has decreased from 25 percent to 20 percent. Just in the last ten years, deaths from heart disease, the biggest killer of Americans, have decreased by 25 percent.

Across every age group, an unheralded revolution has occurred. Check out how relevant these innovations have been to your own life in the following table. The death rate in every age group has dropped by 29 percent or more.

The Silent Miracle Revolution

Annual death rate per 100,000 people in the following age ranges:

Your age	1970	2000	change
0-1	2,142	729	-66 %
1-4	85	33	-61 %
5-14	41	19	-54 %
15-24	128	82	-36 %
25-34	157	108	-31 %
35-44	315	200	-37 %
45-54	730	431	-41 %
55-64	1,659	1,004	-39 %
65-74	3,583	2,428	-32 %
75-84	8,004	5,688	-29 %

Why Did the Greatest Advances in Saving Lives Occur While We Were Lucky Enough to be Alive?

Thanks to science. Life sciences like medicine were slower than the physical sciences to adopt empirical methods, primarily because living systems are so darn complex. It was only after chemistry had bcome a mature science that researchers such as Landsteiner, Florey, Müller, and Endo could move the life sciences so dramatically ahead.

History has demonstrated over and over again that as soon as scientific methods are applied to a field, prodigious new knowledge emerges that replaces the dogma that existed before. Thanks to science, astrology became astronomy, dynamics became physics, and alchemy became chemistry. Science is much more than a collection of knowledge—it is a way of thinking. Specifically, science is a form of reasoning that relies on evidence. Knowledge used to be passed down from generation to generation, resulting in rote memorization of whatever the authorities taught. Science refuses to defer to authority; instead it insists upon demonstrable evidence that allows anyone to look and verify the truth for themselves. For example, instead of believing an authority who insists that Jesus cared more about war than the sick, a scientific approach can be taken by using statistics to count the number of times the sick were mentioned in the gospels as compared to war—the ratio is 29 to 1.

Science revolutionized medicine by insisting that a treatment had to be demonstrable in repeatable experiments before it could be claimed to be a cure.

Medicine without Science—The Death of George Washington

One snowy day in December of 1799, the sixty-eight-year-old George Washington came in from riding his horse around his farm. That evening he complained of a sore throat. Over the next two days, he became hoarse, and began to have trouble breathing. Three doctors saw him, and following the textbooks of the day, each decided the best treatment was to bleed him in the hopes that doing so would stop the inflammation in his throat. At that time, the proponents of bleeding thought the body had twice as much blood as the twelve pints it actually has, so the doctors removed almost half the blood in Washington's body. He died the next evening.

The problem was that Washington's treatment was not based on scientific evidence. Three hundred years ago, about the only double-blind experiments were when a doctor treated both eyes of a patient with the wrong salve. Washington's doctors had learned from other doctors the accepted treatment, and they earnestly practiced what they had been taught, but there had never been any controlled experiments proving that bleeding worked or didn't work. The doctors' reliance on knowledge passed down from authorities prevented them from properly analyzing the problem. The real danger was that Washington's inflamed throat was obstructing his air passage, and most likely he suffocated. Today's prescription of putting a tube down the throat to prevent suffocation was not used successfully until almost 100 years later. But don't miss the point—it is not to blame Washington's doctors, who were performing as they had been trained to do. What is important is to understand why their treatment failed. It wasn't an irresponsible mistake. It was a systemic mistake due to the pre-scientific reasoning the medical community had traditionally used.

Dr. David Eddy, a leading critic of traditional medicine and founder of the evidence-based medicine movement, claims that when he was trained in the 1970s, only 15 percent of what doctors did was backed by scientific evidence. Even today some in the movement claim that doctors use scientific evidence for only 25 percent of their treatments. Some doctors have characterized this as a battle between evidence-based medicine and cookbook medicine, a nice descriptive term for the rote retrieval of treatments out of textbooks or medical school lectures. All of the life-saving advances of the past century were due to this primary characteristic of science: examining evidence.

Unfortunately, making scientific arguments based on evidence does not come naturally to most people. A 1993 study by Deanna Kuhn showed that only 40 percent of adults in America could even make an evidence-based argument. Furthermore, only 20 percent of those without a college education could do so. Using evidence in an argument is something most humans have to learn.

Premature Death

Since we will all die someday, what we are really talking about when we speak of saving lives is stopping premature death. One of the most common responses people have with respect to death is fatalism. Since death is inevitable, it is easy to think that living a full life or living longer is not a choice, but is a matter of genetic or environmental luck. However, the evidence presented in this book of well over one billion lives saved strongly suggests that premature death is not a matter of fate. Mankind, with the power of science, has the ability to reduce mortality from many more causes of premature death, ranging from violence to disease.

With disease, premature death is sometimes defined statistically as dying prior to the age of 65. About one-fifth of all the disease deaths in the United States are premature. In 2003, disease caused 520,405 premature deaths, or about 1,440 a day. Vehicular accidents caused 44,757. Murder caused 16,242.

One reason more people do not celebrate the monumental medical advances in this book is the fear that even if individuals' lives can be saved, the cost to do so will drain society's economic resources. So it is reasonable to ask if saving lives is cost-effective.

Is Spending Money on Healthcare a Good Investment?

The media often create the impression that spending on healthcare will eventually bankrupt the country. A similar concern is that allowing people to live into old age will sap the incomes of young people. If we use scientific reasoning and look at evidence, the complete opposite becomes apparent. As David Cutler, Professor of Economics at Harvard, points out, the reason medical care was 4 percent of a much smaller GDP in 1950, and is 16 percent now, is that doctors couldn't actually treat many conditions in 1950. Now, however, when many diseases can be cured, people quite rationally spend more of their money on medical care.

In order to assess the effect of healthcare on the country's economy, over the past decade several economists have done groundbreaking studies to calculate the value of lives saved economically. Their conclusions, some of which will likely garner a future Nobel Prize, is nothing short of amazing. It is summarized in the book *Measuring the Gains from Medical Research: An Economic Approach*, edited by Kevin M. Murphy and Robert H. Topel, both of the University of Chicago. They found that "improvements to life expectancy alone added about $2.6 trillion per year to national wealth between 1970 and 1998. By comparison, average GDP over this period was about $5.5 trillion." William D. Nordhaus, a Professor of Economics at Yale, found that, "Growth in longevity since 1950 has been as valuable as growth in all other forms of consumption combined."

As hard as it is to believe, the great economic advances over the past three decades may not be primarily the result of the computer revolution or the advent of the Internet. Evidence demonstrates that economic gains due to increased health and lives saved over the last half of the twentieth-century dwarf any other industry's contribution.

These numbers are huge, first because a disease eliminated no longer consumes resources. The United States used to spend hundreds of millions of dollars trying to keep people with smallpox from coming into the country. Now that smallpox no longer exists, such spending is no longer necessary. Secondly, preventative measures such as vaccines abolish disease. If millions of people never get a disease like measles, no money is spent trying to cure it. Thirdly, and most importantly, every single person whose life is saved continues to produce and consume. Think of an infant saved. He or she may grow up to be another Steve Jobs and run Apple, one of the most innovative companies in the world.

A simple statistic demonstrates the effect of saving lives on the economy. In 1960, when the population of the United States was about 200 million, the number of deaths each year was 2.7 million. In 2000, a comparable population of 200 million, adjusted for a similar age distribution, had 1.7 million deaths, a million less. Over decades, the numbers of lives saved add up to the tens of millions, each of whom actively participates in the economy, producing goods and services and paying taxes. Economists also recognize a multiplier effect. Every dollar produced or consumed by a person whose life has been saved affects other people, who sell that person goods, and who then also produce and consume more, which causes other people to produce and consume more, and so on. In addition, there are intangibles. Would globalization, considered by many to be the economic story of the decade, even be possible without the defeat of smallpox, allowing unrestricted travel throughout the world?

Increasing longevity is the greatest economic tool our culture has produced. No other investment has had such high rates of return. Yet most of what the media report is that we have a financial crisis in healthcare. They are right; we do have a financial crisis. From an economic point of view, we aren't spending too much on healthcare—we are investing way too little!

What Can Scientific Evidence Teach Us about Investing in Research to Save Lives?

We live in a nation where we each have a say in how our resources are used. As a society, we can decide if we build huge monuments like those on Easter Island, or if we add more and more square footage to our homes. One of our choices is how much to spend on health-care research. Charity accomplishes much, but it can play only a small role in health-care research because it is so expensive. In 2004, charity contributed about 2 percent of the $101 billion spent in the United States on health-care research, and half of that 2 percent came from the Bill and Melinda Gates Foundation. The 60 percent of all health-care research funding that corporations—primarily the drug companies—contribute is significant. But corporations only spend money on big problems, where there are big profits to be realized, and on relatively focused problems, where a profitable solution can be achieved relatively quickly. Funding for basic, untargeted research, and research on less common health problems, must come from the government. And as much as some people are loath to contribute, the government gets its money from taxes. Are we basing our decisions on how to spend our tax money using evidence, as science has taught us? If we are, what should we choose to spend our tax money on, if we want to save lives?

Taxes Spent on Researching How to Save Lives

2004 Federal Government Research to Save Lives

$69 Billion	Defense
$38 Billion	Disease
$58 Million	Vehicle Safety

To be clear, $69 billion is not the defense budget, which for 2004 totaled $380 billion and for 2008 will total $481 billion (not counting what is spent on the wars in Iraq and Afghanistan). The figure we are discussing is only the defense *research* budget. The budgets have changed since 2004. In 2008, the outlay for defense research increased to $81 billion, while federal funding for health-care research was unchanged.

If this spending were rational, it would be based on the evidence of need. Here are the numbers of deaths caused by each of the above categories.

Deaths by Category Per Year

520,405	Disease prior to age 65 in 2003
44,757	Vehicle Accidents in 2003
277	Yearly average war and terrorist deaths for the past thirty years, thru 2007.

We can calculate the amount spent on each individual life:

2004 Federal Research Money Spent Per Life Lost by

War and terrorism	$249,000,000
Disease prior to age 65	$73,000
Vehicular Accidents	$1,300

Let's use a military analogy: medical triage. If a MASH unit with 3,401 doctors is confronted with 1,879 wounded soldiers, how would they allocate the doctors? Would they assign 3,400 to one wounded soldier and one to all the rest? That is the way we are allocating our research funds. For every war death, there are 1,878 disease deaths, yet we are spending 3,400 times as much on preventing a war or terrorism death as we are on preventing a death, perhaps your child's, from disease. And we are spending 190,000 times as much to prevent a war or terrorism death as we are to prevent a traffic death.

Our Declaration of Independence states that it is a self-evident truth that all people are created equal. The Golden Rule implies that everyone should be treated equally. Obviously, our government officials do not follow either in allocating research funding.

Some people would say that war may not be the biggest threat to life now, but historically it has been. That argument ignores the evidence. In the twentieth century, one-and-a-half times more people died from one disease—smallpox—than died in all the wars in all the countries combined (300 million to 188 million, and the war deaths include combat and noncombat deaths, as well as those resulting from politically caused famines and genocide). Every single day in the United States, 1,440 non-elderly people die prematurely. That is one per minute, like a 9/11 terrorism attack every two days. Is it because there are no television cameras on their pale, sad faces that we think less about these than about those who die in war? Is it because they die with less drama?

Still believe war is the gravest danger? More than twice as many people have died in vehicular accidents in the United States than have died in all the wars the United States has ever fought. (The highest count of war deaths in all of the wars in U.S. history, including both combat and non-combat deaths, is 1,541,650. Total vehicle deaths through 2005 are 3,393,806). Don't think health insurers should cover mental health?

More people have died from suicide since 1940 than have died in all the wars the United States has ever fought. Against providing health insurance to those who can't afford it? More people die each year due to having no health insurance than are murdered. Believe health-care providers should cut corners to save money? More than two times as many people die each year from medical errors than are murdered.

The evidence indicates that we are not spending our resources on researching what is really killing Americans! Let's do a thought experiment. Let's imagine that you and your spouse are going to have a baby and you have a certain amount of resources in the form of taxes that you are willing to pay and you want to use those resources to protect your child from deadly harm over his or her lifetime. First you will determine what your child will most likely die from.

Probability of Your Child Dying

Before the age of 65 due to disease	1/6
Lifetime due to auto accident	1/84
Lifetime due to suicide	1/116
Lifetime due to murder	1/217
Lifetime due to war or terrorism	1/13,000

Your child is 2,000 times more likely to die prematurely from disease than from war or terrorism, and more than 100 times more likely to die from an auto accident or suicide than from war or terrorism. It is our choice, how we spend our taxes for research.

The Irony Is that the United States Is the Best in the World at Research!

One day you may be in a hospital, holding the hand of a loved one. He or she will die. You will walk out of the hospital alone, tears streaming down your cheeks, your loved one dead. You will get into your big, new shiny car and drive home to your house, which is $100,000 more expensive than the one you bought in 1990. You will walk through the house, not wanting to sit down, alone without your loved one, and you will find yourself in a back bedroom. You paid extra for the back bedroom, for more carpet and drywall. You chose to spend your money on the fancier car, the bigger house, rather than pay more taxes. And now you stand alone in a room no one will live in. Was your choice made using scientific evidence?

This is not a harangue to make you wring your hands and lament that the United States of America can never do anything right. What does the evidence say about the United States? Based on a Chinese university's study, the United States has seventeen of the top twenty universities in the world. The United States leads the world in money spent on healthcare research by private firms. The United States leads the world in health-care research funded by the government. Europe doesn't come close, nor does any other country. The truth is that the U.S. has done more right than any other country in the history of the world—ever! We excel. Medical research is a stream of successes. Half of our ten heroes were born and educated in the U.S., and the whole world owes them an enormous debt of gratitude. And while globalization may take manufacturing jobs or even service jobs overseas, our nation is far out front in scientific research. We can be the nation that cures the world. We can be the nation that all others look to, to heal their loved ones, to model their research spending after. We can be the nation that teaches the rest of the world that scientific evidence is the answer to mankind's problems.

Scientific evidence indicates that we could save tens of thousands more lives with a few policy changes.

Policy Change No. 1—Cut Vehicular Deaths to 10,000 by the Year 2015

A NASCAR vehicle flips thirteen times and the driver gets out grinning from ear to ear. We have the technology and scientific intelligence to drastically cut the probability of death in an automobile accident to much less than 1/6,500 per person each year. How can a parent in his or her right mind spend money for GPS systems and engines with much higher horsepower than necessary when their child's car is not designed to be safe?

Policy Change No. 2—Reform Health Insurance

Studies have shown that 20,000 non-elderly people a year die from not having health insurance. Those without health insurance often delay going to the doctor, even in emergencies, due to concern about the cost of treatment. For a nation as wealthy as ours to allow 20,000 people a year to die from not having health insurance is shameful. Perhaps a new way is needed to help wealthy people in power empathize with the problem the working poor have in paying for health insurance. Since everyone who has health insurance pays for the treatment of those without health insurance either through taxes or higher insurance premiums, health insurance, even if privately purchased, can be viewed as a tax. If health insurance is viewed as a tax, then for the forty percent of workers whose employers do not provide health insurance, it is a draconian tax. A health insurance policy for an average family of four now costs over $10,000 a year. This equals a 52 percent tax on a household at the 20th percentile of income and a 20 percent tax on a household at the 50th percentile. This is why one-in-three non-elderly Americans go periods of time with no health insurance.

Policy Change No. 3—Enhance the Jobs of Those Who Provide Medical Care

When we have a sick child or loved one, where would we be without those who provide us with healthcare? Upon visiting any doctor's office or hospital, it is immediately apparent that health-care providers work hard and care about those who are ill. In fact, the average doctor works fifty hours a week. Evidence indicates that health-care professionals do more for the economy than capitalists, yet they pay much higher taxes. Doctors currently pay a top rate of 35 percent on their income. The top rate capitalists pay is 15 percent. Economists have shown that health-care professionals and researchers are more important for the economic success of the country than capitalists, so they should have a top income tax rate no higher than the capitalists' 15 percent rate. By reducing their taxes, doctors can take on fewer patients and still make as much money as they do now, thereby using their extra time to provide better care to their patients and to study new advancements in medicine. With a reduction in their taxes, more of our brightest students will be drawn to medicine and research, which could solve the looming shortage of doctors and other health-care professionals that is expected in the coming decades, as well as provide more research firepower to find new ways to save lives.

Primary Priority—Increase Health-care Research to $153 Billion

A drug company's CEO who refused to invest a portion of his company's profits in researching and developing new drugs would be considered irresponsible and fired. Government should be held to a similar standard. What have our government officials done? The 2008 federal budget increased defense research funding to $81 billion, an increase of 17 percent since 2004. The primary medical research agencies in the government are the National Institutes of Health. The NIH's budget has not even kept up with inflation over the past four years. Evidently, our government officials care a lot more about creating weapons that kill people than creating weapons that save people.

Kenneth G. Manton and his colleagues at Duke University have analyzed the amount of medical research the federal government should be funding, taking into account the contribution of past medical advances obtained through research, the economic benefit of such research, the aging of our population, and other factors. In a paper published in 2007, they suggest that the federal government should be spending $153 billion a year on health research, which is four times its current amount. If enacted, this would increase spending to almost $300,000 per disease death, still 1,000 times less than that spent on each war/terror death. Currently, less than 20 percent of scientists who request a grant to study disease are being funded. How many scientists greater than Einstein are not being funded? Will your child have a disease that could have been studied by one of the 80 percent of grants that are now denied funding?

Buttonhole Your Congressperson

The evidence is clear and there is ample academic research to prove that our nation needs to drastically change our research priorities. In the end, however, there is only one person who can ultimately decide to save more lives: your congressperson. To be sure, entrenched lobbyists will fight any increase in medical research funding. Your congressperson is paid by these lobbyists to listen to them and ignore you, but he or she also knows that if enough people care about something, any congressperson can be removed from office. So shake your congressperson's hand, look him or her in the eye, and with your other hand give them a picture of a loved one who has died prematurely, or one who is alive and whom you don't want to die early. Tell your congressperson: "Only you can prevent premature death."

—Billy Woodward

Determining the Number of Lives Saved

by Amy R. Pearce, Ph.D.

An approximate answer to the right problem is worth a good deal more than an exact answer to an approximate problem.
— John Tukey, Ph.D. and statistician

As THIS project's statistician, my task was to tabulate figures that were in all practicality impossible to pinpoint. Consequently, each number represents a conservative estimate of lives saved for each of the "life savers," based on the best available data and a common-sense approach.

Whenever possible, I employed a consistent and logical multistep process. I used both aggregated and disaggregated world population figures from the twentieth century and beyond to count the potential beneficiaries from each discovery, and I established a timeline from the discovery's invention and first widespread use through 2008. I calculated lives saved based on published data from the most reputable, corroborative, and accessible evidence and resources (WHO, CDC, NIH, UNICEF, and databases such as Medline, Lexus Nexus, and PubMed). Yet, on occasion, these yielded insufficient data and I then considered alternate sources such as Internet websites and a variety of nonacademic literature. Software programs such as Excel and SPSS were used for calculations and graphics.

I found population data for countries with ready access to the then-new cure, and also mortality and morbidity statistics for targeted illnesses, then determined a percentage of this population that would logically have died without the intervention. For the most part, the information for the developed world (North America, Western Europe, parts of South America, Southeast Asia, and certain other countries) was the most accessible and reliable, and so the majority of calculations were extrapolated from these figures.

Statistics from the developing world were scarce, and if figures were not available, or could not be credibly inferred, they were not included in extrapolated figures. Hence, while many lives-saved totals are astoundingly high, I believe each result is wholly conservative and likely underestimates the true number of beneficiaries.

Chapter 1: Al Sommer—Over 6 Million Lives Saved
The Eye Doctor Who Discovered a Better Use for Vitamin A

The vitamin A supplement campaign was launched globally in the 1990s, and UNICEF tracks the number of people who have received this intervention. I was able to make a relatively solid inference because the numbers are fairly stable, averaging between 300,000–500,000 lives saved yearly since the campaign was initiated. Over 100 million young children suffer from vitamin A deficiency and supplements are given biannually, mostly to children 0–5 years of age. Vitamin A deficiency is a contributing factor in deaths each year from both diarrhea and measles among children under five years of age.

Chapter 2: Akira Endo—Over 5 Million Lives Saved
Statins: Life Extension for the Baby Boomers

This one was a struggle. A relatively new drug class, statins became widely recognized in the early 1990s and have since been progressively used by older patients. Concerning people over the age of 20 in the U.S. (~74 percent of the population), in 1993 approximately 3 percent were on a statin; by 2006, nearly 10 percent of this population was taking a statin. To determine total potential beneficiaries, I found annual U.S. population data from 1993–2008, totaled 74 percent of this figure, and then calculated 3 percent of this total for roughly the first few years, with an incremental increase over the years up to 10 percent for 2008. This figure was extrapolated to include all developed countries with similar lifestyles, health-care systems, and availability and use of statins. Interestingly, although statins have been introduced in a number of other large populations (China, India, etc.), their effects were less dramatic due to differences in genetics, diet, and lifestyles and were therefore not included in our final total. The data came from a large-scale empirical study in Europe published in 1995 known as the "Scandinavian Simvastatin Survival Study," or 4S, which estimated that statins reduced the risk of death by 30 percent among people who have coronary heart disease. While there is controversy over the benefits of statins to people with high cholesterol but a low risk of having a heart attack, there is ample proof of their benefit to those with high risk factors.

Chapter 3: Bill Foege—Over 122 Million Lives Saved
The Eradication of Smallpox

Smallpox was an exceptional killer, even though by the time the smallpox eradication effort began Jenner and others had been vaccinating for some 170 years and the U.S. was smallpox-free. The figure is an estimate generated from 1967–2008, based on the percentage of the world population at risk, using WHO reports of well over 2 million deaths per year prior to global eradication (smallpox was endemic in thirty-three countries at this time). This is quite a conservative estimate.

Chapter 4: David Nalin—Over 50 Million Lives Saved
ORT: A Revolutionary Therapy for Diarrhea

Dr. Nathaniel F. Pierce estimated that between 1979 and 1990, diarrhea deaths in children declined from 4.6 million to 2.9–3.3 million annually, and from 1991 to 1999 declined further to 1.5 million annually, while ORT became available in 110 countries. These figures corroborate a *British Medical Journal* article in 2007 that says over 50 million lives have been saved by ORT.

Chapter 5: Norman Borlaug—
Over 245 Million Lives Saved
A Green Revolution to Enhance Nutrition

There are numerous published claims that Norman Borlaug, through the Green Revolution, saved over a billion lives, but there seem to be no sources of statistics backing up these statements. They seemed to have arisen due to predicted apocalyptic famines. The famines were unlikely to have occurred on the scale predicted with or without the Green Revolution. However, this does not lessen the actual impact the Green Revolution had on mortality. Numerous studies have shown that childhood nutrition has an emphatic effect on childhood mortality and longevity. Gabriel Popkin, who did much of the background research on this topic, compiled statistics on the changes the Green Revolution brought to nutritional levels in children in parts of the world most affected by the Green Revolution, and its effect on childhood mortality.

Chapter 6: John Enders—
Over 114 Million Lives Saved
The Father of Modern Vaccines

Before the availability of the live measles vaccine, an estimated 6–8 million children died of measles annually throughout the world. That number has since been reduced to less than 1 million. In the United States, the incidence of measles has been reduced to fewer than 100 cases annually, and virtually all these cases can be traced to importation of the measles virus from abroad. Today, there are many other countries in the Western Hemisphere where indigenous transmission of measles no longer occurs. To determine the number of lives saved, I took the estimated annual deaths in the prevaccine era and distributed the percent reduction across twenty-one years for the period of 1977 to 2008.

Regarding the polio vaccine and the over 1.3 million lives it saved, I used mostly CDC and WHO reports to determine two estimates of lives saved between the years of 1960 and 2008 for the entire developed World. These both resulted in very close final tallies, so I felt fairly confident they were among the best estimates I could make with the information available. I then used these data to extrapolate to the rest of the world. Although India accounts for about +70 percent of all reported cases (and some regions within Africa, especially Nigeria, are also high), eradication efforts really only began there full scale in the past fifteen years, with strong results in the late 1990s and early 2000s, so it was difficult to put a number on the rest of the world. However, I took a percentage for the world from 2000–2008, when the efforts of global eradication became more fully realized.

Chapter 7: Paul Müller—Over 21 Million Lives Saved
DDT and the Prevention of Malaria

Getting this number presented another lesson in the value of using original sources. An early lead on DDT put lives saved at the 500 million mark, and this number was touted by several reputable sources. However, in tracking down the origin of this number it appeared to come from an article by Clive Shiff who quoted "500 million freed from the threat of disease," not exactly lives saved. I revisited DDT and ran two different calculations with amazingly close results. They both estimated lives saved by DDT in the 20 millions. One estimate was based on 1950 world population demographics and deaths from malarial disease in 1000s per year for three different age groups. There were just over 2 million deaths annually from malaria in 1950 and this dropped drastically over the next decade with the use of DDT, so I took this number and multiplied by 10 to make a conservative estimate for the years 1955–1964. The second estimate was based on world population from 1950 to 1965 and annual deaths from malaria. Estimates put it at 1740 per million in 1930, to 592 in 1940, 480 in 1950, and down to 1 per million in 1960, when DDT had been globally used in eradication programs. The numbers jump up in the 1970s when DDT was largely taken out of the eradication programs, and after resistance had built up due to overuse in agricultural applications. DDT has since begun to be used again in parts of Africa, where malaria remains a scourge, but I did not include this timeframe since it began too recently to generate any solid statistics.

Chapter 8: Howard Florey—
Over 80 Million Lives Saved
Penicillin: The Miracle of Antibiotics

This was a truly impossible task because penicillin is prescribed to so many people in so many forms for so many things, some life-threatening, others not. Hailed as a miracle drug when introduced during WWII, penicillin continues to fight infection around the world, and the number of lives saved offered here probably just scratches the surface. One Swedish source cited 200 million lives saved, but without explication. Although widely used in the developing world, there is virtually no accessible data. This statistic includes penicillin as used for treating sepsis, bacterial meningitis, syphilis, and pneumonia, but does not include treatments for other maladies, nor the use of penicillin derivatives.

Chapter 9: Frederick Banting—
Over 16 Million Lives Saved
Insulin: The First True Miracle Drug

According to data published by the U.S. Census Bureau, International Diabetes Federation, and the National Diabetes Surveillance System, between 10 and 20 million people worldwide are believed to have Type 1 diabetes. Using population growth estimates from 1950–2008 and an intermediate incidence rate of 6/100,000 new cases per year, over 16 million lives have been saved. I made no effort to determine the number of lives saved for type 2 diabetes (formerly adult onset diabetes), although insulin definitely saves some lives from it as well.

Chapter 10: Karl Landsteiner—
Over 1 Billion and 38 Million Lives Saved
The Superman Scientist: Discoverer of Blood Groups

Every source quoted an amazing number of transfusions and potential lives saved in countries and regions worldwide. High impact years began around 1955, and I based my calculations loosely on a ratio of one life saved per 2.7 units of blood transfused. In the U.S. alone, blood transfusions save an estimated 4.5 million lives each year. From these data, I determined that 1.5 percent of the population was saved annually by blood transfusions and I applied this percentage on population data from 1950–2008 for North America, Europe, Australia, New Zealand, and parts of Asia and Africa. This rate may inflate the effectiveness of transfusions in the early decades but excludes the developing world entirely. Since the late 1980s, blood donations have declined and the surplus of blood that has existed will soon end. Call this efficiency, but there's also a risk of future transfusion demands not being met. If your blood type is 0 negative (like me) your phone line is probably hot with calls from the local Red Cross!

Sources

Chapter 1. Al Sommer:
The Eye Doctor Who Discovered a Better Use for Vitamin A
Written by Billy Woodward and Joel Shurkin

Avery, Mary Ellen. "An Interview with Alfred Sommer." Lasker Foundation. 1997. http://www.laskerfoundation.org/awards/1997_c_interview_sommer.html (accessed June, 23, 2007).

Hussaini, G., I. T. Tarwotjo, and A. Sommer. "Cure for Night Blindness." Letter to editor. *American Journal of Clinical Nutrition* 31, no. 9 (September 1978): 1489.

Keusch, G. T. "Vitamin A Supplements—Too Good Not to Be True." *New England Journal of Medicine* 323, no. 14 (October 4, 1990): 985–986.

Lasker Luminaries. "Alfred Sommer." Lasker Foundation. 2002. http://www.laskerfoundation.org/awards/pdf/1997_sommer.pdf.

Mosley, W. H., K. Bart, and A. Sommer. "An Epidemiological Assessment of Cholera Control Programs in Rural East Pakistan." *International Journal of Epidemiology* 1 (1972): 5–11.

Simpson, Brian W. "The Other Al Sommer." John Hopkins Public Health. 2005. http://magazine.jhsph.edu/2005/fall/prologues/ (accessed June 23, 2007).

Sommer, A. Interview. Baltimore, MD, July 31, 2007.

———. "Effectiveness of Surveillance and Containment in Urban Epidemic Control" (Concerning the 1972 smallpox outbreak in Khulna municipality, Bangladesh). *Amercian Journal of Epidemiology* 99 (1974): 303–313.

———. *Nutritional Blindness: Xerophthalmia and Keratomalcia.* New York: Oxford University Press, 1982.

———. "Mortality Associated with Mild, Untreated Xerophthalmia." Thesis. *Trans Am Ophthalmol Soc* 81 (1983): 825–853.

——. "Vitamin A Deficiency and Its Consequences: A Field Guide to Detection and Control." World Health Organization. 1995.

——. "A Bridge Too Near. The Progress of Nations—Nutrition." 1995. http://www.whale.to/v/sommer.html (accessed June 28, 2007).

——. "Rx for Survival: A Global Challenge." Public Broadcasting System. 2006. http://www.pbs.org/wgbh/rxforsurvival/series/champions/alfred_sommer.html.

——, G. Hussaini, I. Tarwotjo, J. Susanto, and J. S. Saroso. "History of Nightblindness: A Simple Tool for Xerophthalmia Screening." American Journal of Clinical Nutrition 33 (1980): 887–891.

——, M. Khan, and W. H. Mosley. "Efficacy of Vaccination of Family Contacts of Cholera Cases." *The Lancet* 1 (1973): 1230–1232.

——, and W. H. Mosley. "East Bengel Cyclone of November 1970: Epidemiological Approach to Disaster Assessment." *The Lancet* 1 (1972): 1029–1036.

——, and W. H. Mosley. "Ineffectiveness of Cholera Vaccination as an Epidemic Control Measure." *The Lancet* 1 (1973): 1232–1235.

——, and M. Loewenstein. "Nutritional Status and Mortality: A Prospective Validation of the QUAC Stick." *American Journal of Clinical Nutrition* 28 (1975): 287–292.

——, I. Tarwotjo, G. Hussaini, and D. Susanto. "Increased Mortality in Children with Mild Vitamin A Deficiency." *The Lancet* 2 (1983): 585–588.

——, I. Tarwotjo, Muhilal, E. Djunaedi, and J. Glover. "Oral Versus Intramuscular Vitamin A in the Treatment of Xerophthalmia." *The Lancet* 315, no. 8168 (March 1980): 557–559.

——, I. Tarwotjo, E. Djunaedi, K. P. West, A. A. Loedin, R. Tilden, L. Mele, and the Aceh Study Group. "Impact of Vitamin A Supplementation on Childhood Mortality: A Randomised Controlled Community Trial." *The Lancet* 1 (1986): 1169–1173.

——, K. P. West, Jr. *Vitamin A Dificiency—Health, Survival and Vision*. New York: Oxford University Press, 1996.

——, and Keith P. West. "Delivery of Oral Doses of Vitamin A to Prevent Vitamin A Deficiency and Nutritional Blindness." Center for Epidemiologic and Preventive Opthalmology. 1987.

——, and W. E. Woodward. "The Influence of Protected Water Supplies on the Spread of Classical/Inaba and El Tor/Ogawa Cholera in Rural East Bengal." The Lancet 2 (1972): 985–987.

Tarwotjo, I., A. Sommer, T. Soegiharto, D. Susanto, and Muhilal. "Dietary Practices and Xerophthalmia among Indonesian Children." *American Journal of Clinical Nutrition* 35, no. 3, (March 1982): 574–581.

Tielsch, J. M., and A. Sommer. "The Epidemiology of Vitamin A Deficiency and Xerophthalmia." *Annual Revue of Nutrition* 4 (1984): 183–205.

Villamor, E., and W. W. Fawzi. "Effects of Vitamin A Supplementation on Immune Responses and Correlation with Clinical Outcomes." *Clinical Microbiology Revue* 18, no. 3, (July 2005): 446–464.

Chapter 2: Akira Endo:
Statins—Life Extension for the Baby Boomers
Written by Billy Woodward and Joel Shurkin

Brown, M. S., and J. L. Goldstein. "A Tribute to Akira Endo, Discoverer of a 'Penicillin' for Cholesterol." *Atherosclerosis Supplements* 5 (2004): 13–16.

Brown, M. S., and J. L. Goldstein. "Lowering Plasma Cholesterol by Raising Ldl Receptors." *New England Journal of Medicine* 305, no. 9, (1981): 515–517.

Brown, M. S., and J. L. Goldstein. Brown-Goldstein Laboratory. UT Southwestern Medical Center. http://www.utsouthwestern.edu/utsw/cda/dept14857/files/114532.html (accessed October 29, 2007).

Brown, M. S. and J. L. Goldstein. "A Receptor-Mediated Pathway for Cholesterol Homeostasis." Nobel lecture. The Nobel Assembly at the Karolinska Institute. 1985. http://nobelprize.org/nobel_prizes/medicine/laureates/1985/brown-goldstein-lecture.pdf.

Charatan, Fred. "Severity of Heart Attacks in United States May Be Declining." *British Medical Journal* 318 (1999): 896.

Endo, Akira. E-mail interview, October 22, 2007.

——. *Designed by Nature—The Birth of the Greatest New Medicine in History*. Translated by Hiroko Hollisj. Tokyo: Medical Review Company, Ltd. (2006)

———. *The Dicovery of the New Medicine Statin—A Challenge to Cholesterol.* Translated by Hiroko Hollis. Tokyo: Iwanami Shoten, 2006.

———. "Monacolin K: A New Type of Cholesterolemic Agent Produced by a Monoacus Species." *Journal of Antibiotics* 32, no. 8, (1979): 852–854.

———. "The Discovery and Development of Hmg-Coa." *Journal of Lipid Resea*rch 33 (1992): 1569–1578.

———. "I Finally Tested Statin on Myself!" *Artherosclerosis Supplements* 5 (2004): 31.

———. "The Origin of the Statins." *Atherosclerosis Supplements* 5, no. 30, (2004): 125–130.

———, M. Kuroda, and Y. Tsuita. "New Inhibitors of Cholesterogenesis Produced by Penicillium Citrinum." *The Journal of Antibiotics* 31, no. 12, (1976): 1346–1368.

———, M. Kuroda, and K. Tanzawa. "Competitive Inhibition of 3-Hydroxy-3-Methylglutaryl Coenzyme: A Reductase by M1-236a and M1-236b Fungal Metabolites, Having Hypocholesterolemic Activity." *FEBS Letters* 72, no. 2, (1976): 323–325.

King, M. W. "Introduction to Cholesterol Metabolism." The Medical Biochemistry Page. Copyright 1996-2006. http://themedicalbiochemistrypage.org/Cholesterol.html#introduction (accessed November 3, 2007).

Landers, Peter. "Drug Industry's Big Push into Technology Falls Short." *Wall Street Journal*, February 24, 2004.

———. "How One Scientist Intrigued by Molds Found Statin." *Wall Street Journal*, January 9, 2006.

Nogrady, B. "All in the Genes." Austrailian Doctor.com Top 50 Medical Innovations. http://www.australiandoctor.com.au/Common/ContentManagement/AusDoc/pdf/top 50_pdfs.pdf (accessed November 3, 2007).

Nakamura, Jazuo. "A Unique Cholesterol-Lowering Agent That No One Has Ever Had Before." *Atherosclerosis Supplements* 5 (2004): 19–20.

Oliver, Michael, Philip Poole-Wilson, and et al. "Lower Patients' Cholesterol Now." *British Medical Journal* 310 (1995): 1280–1281.

Olsson, Anders G. "The Importance of the Emergence of Statins to a Lipodologist-Clinician: A Personal Perspective." *Atherosclerosis Supplements* 5 (2004): 27–28.

Pedersen, Terje. "Randomised Trial of Cholesterol Lowering 4,444 Patients with Coronary Heart Disease: The Scandanavian Simvastatin Survival Study." *The Lancet* 344 (1994): 1383–1387.

Rosmond, Waynes, and et al. "Heart Disease and Stroke Statistics—2007 Update." *Circulation* 115 (2006): 69–71.

Shepherd, J., et al. "Prevention of Coronary Heart Disease with Prevastatin in Men with Hypercholesterolemia." West of Scotland Coronary Preventions Study Group. *New England Journal of Medicine* 333, no. 20, (1995): 1301–1307.

Thompson, Gilbert R. "Unpremeditated Tribute to Akira Endo." *Atherosclerosis Supplements* 5 (2004): 29.

Winslow, Ron. "Studies Point to Drop in Heart Attacks, Cholesterol." *Wall Street Journal*, October 12, 2005.

Yamamoto, Akira. "Determination to Treat Patients with Fh— How Fundamental Techniques in Medical Consultation Saved the Day." *Atherosclerosis Supplements* 5 (2004): 25–26.

Yamamoto, A., H. Mabuchi, and et al. "Thirty Years of Statins Festschrift in Honor of Dr. Akira Endo." *Atherosclerosis Supplements* 5, no. 3, (2004): 1–130.

Yamauchi, Kimiko. *The Most Acknowledged Medicine in the World. Tokyo: Shogakukan*, 2007 (translated by Hiroko Hollis).

———. "The Nobel Prize in Physiology or Medicine 1985." The Nobel Assembly at the Karolinska Institute. 1985. http://nobelprize.org/ nobel_prizes/medicine/laureates/1985/press.html.

———. "Cholesterol." *Encyclopedia Britannica* 2006 Ultimate Reference Suite DVD. 2006.

———. "About Cholesterol." American Heart Association. 2007. http:// www.americanheart.org/presenter.html?identifier=512.

———. "Ateriosclerosis." *Encyclopedia Britannica* 2006 Ultimate Reference Suite DVD. 2007.

———. "Know the Facts, Get the Stats." American Heart Association. 2007. http://www.americanheart.org/downloadable/heart/ 116861545709855-1041%20KnowTheFactsStats07_loRes.pdf.

Chapter 3: Bill Foege:
The Eradication of Smallpox
Written by Billy Woodward and Joel Shurkin

Dubos, Rene. *Man Adapting*. New Haven: Yale University Press, 1980.

Falola, Toyin. *Culture and Customs of Nigeria*. Westport: Greenwood Press, 2001.

Foege, W. H., S. O. Foster, and J. A. Goldstein. "Current Status of Global Smallpox Eradication." *American Journal of Epidemiology* 93, no. 4, (1971): 223–233.

——, J. D. Millar, and J. M. Lane. "Selective Epidemiologic Control in Smallpox Eradication." *American Journal of Epidemiology* 94, no. 4, (1971): 311–315.

——, J. D. Millar, and D. A. Henderson. "Smallpox Eradication in West and Central Africa." *Bulletin of the World Health Organization* 52 (1975): 209–222.

——. "Commentary: Smallpox Eradication in West and Central Aftrica Revisited." *Bulletin of the World Health Organization* 76, no. 3, (1998): 233–235.

——. "The Wonder That Is Global Health." Peacescorpsonline.org. 2003. http://corpsonline.org/messages/messages/467/1011931. html.

——. Speech. World Affairs Dinner. 2005. http://74.125.95.104/ search?q=cache:4xUaM3FCIVwJ:www.world-affairs.org/ documents/FoegeSpeech.pdf+world-affairs.org/documents/ FoegeSpeech&hl=en&ct=clnk&cd=1&1=us&client=firefox-a.

——. Phone interview. February 25, 2008.

——. *The Anatomy of Smallpox Eradication in India: Personal Narrative*. Unpublished manuscript, 2008.

——. Acceptance remarks—The 2001 Public Service Award. 2001. http://www.laskerfoundation.org/awards/2001_p_accept_foege. html.

Glynn, I., and J. Glynn. *The Life and Death of Smallpox*. New York: Cambridge University Press, 2004.

Griffin, T. "The Man Who Helped Banish Smallpox from the Earth Is the 1994 Alumnus of the Year." 1994. www.washington.edu/ alumni/columns/top10/calling_the_shots.html.

Loftus, M. J. "Health for All." *Emory Magazine*. 2002. http://www. emory.edu/EMORY_MAGAZINE/winter2002/foege.html.

McCarthy, J. D., and W. H. Foege. "Status of Eradication of Smallpox (and Control of Measles) in West and Central Africa." *Journal of Infectious Di*seases 120, no. 6, (1969): 725–732.

Rosenfield, A. "An Interview with William Foege." 2001. http:// www.laskerfoundation.org/awards/2001_p_interview_foege. html.

Schweitzer, A. "Teaching Reverence for Life." http://www.salsa.net/ peace/conv/8weekconv1-6.html.

Shurkin, Joel N. *The Invisible Fire: The Story of Mankind's Victory over the Ancient Scourge of Smallpox*. New York: G. Putnam and Sons, 1979.

Spector, M. "What Money Can Buy." *The New Yorker*, October 24, 2005.

Stolber, Sheryl Gay. "A Nation Challenged: Public Health." *The New York Times*, September 30, 2001.

Thompson, D. March e-mail interview, 2008.

Tucker, J. B. *Scourge: The Once and Future Threat of Smallpox*. New York: Grove Press, 2001.

Smallpox Disease Overview." Centers for Disease Control and Prevention. 2004. http://www.bt.cdc.gov/agent/smallpox/overview/ disease-facts.asp

"History and Epidemiology of Global Smallpox Eradication." Centers for Disease Control and Prevention. http://www.bt.cdc.gove/ agent/smallpox/training/overview/pdf/eradicationhistory.pdf.

Chapter 4: David Nalin:
ORT—A Revolutionary Therapy for Diarrhea
Written by Billy Woodward and Joel Shurkin

"Work Session 4: Malabsorption/Ion Channels." University of Leeds School of Medicine. 2007. http://www.bmb.leeds.ac.uk/teaching/ icu3/index.html (accessed November 4, 2007).

Cash, R. A., D. R. Nalin, R. L. Rochat, B. Reller, Z. A. Haque, and Rahman, A. S. M. M. "A Clinical Trail of Oral Therapy in a Rural Cholera-Treatment Center." *American Journal of Tropical Medicine* 19, no. 4, (1970): 653–656.

Elliot, J. "A Life-Changing Experience." *Albany Medical College Alumni Bulletin*, 2007.

Farthing, M. J. G. "History of ORT." *Drugs* 36 (supplement 4) (2007): 80–90.

Foex, B. A. "How the Cholera Epidemic of 1832 Resulted in a New Technique for Fluid Resuscitation." *Emergency Medical Journal* 20 (2003): 316–318.

Fontaine, O., P. Garner, and M. K. Bhan. "Oral Rehydration Therapy: The Simple Solution for Saving Lives." *British Journal of Medicine* 334 (suppl 1) (2007): 14.

Guerrant, R. L., B. A. Carneiro-Fiho, and R. Dillingham. "Cholera, Diarrhea, and Oral Rehydration Therapy: Triumph and Indictment." *Clinical Infectious Di*sease 37 (2003): 398–405.

Hirschhorn, N. Speech at the Charles A. Dana Awards for Pioneering Achievements in Health and Education. 1990.

Horn, R., A. Perry, and S. Robinson. "Diarrhoea: Why Is a Simple and Inexpensive Treatment Not More Widely Used?" IRC International Water and Sanitation Center. 2006. www.irc.nl/page/31514 (accessed October 2, 2007).

Mendler, J. "Take the Science to the Problem! Oral Rehydration Salt Solution Solves One of Humanity's Most Dire Problems." The Concord Consortium. 2007. www.Concord.org (accessed October 1, 2007).

Nalin, D. R. Joel Shurkin's interview. 2007.

Nalin, D. R. Personal e-mail communication. October 10, 2007.

Nalin, D. R., R. A. Cash, R. Islam, J. Molla, and R. A. Phillips. "Oral Maintenance Therapy for Cholera in Adults." The Lancet 292 (1968): 370–375.

Nalin, D. R., and R. A. Cash. "Oral or Nasogastric Maintenance Therapy for Diarrhoea of Unknown Etiology Resembling Clolera." *Trans. R. Soc. Trop. Med.* HI-g 64, no. 5, (1970): 769.

Parashar, U. D., C. J. Gibson, J. S. Bresee, and R. I. Glass. "Rotavirus and Severe Childhood Diarrhea." Emerging Infectious Diseases [serial on the Internet]. 2006. http://www.cdc.gov/ncidod/EID/vol-12no02/05-0006.html (accessed October 28, 2007).

Quotah, E. "A Not-So-Simple Solution." *Harvard Public Health Review* (2006).

Ruxin, J. N. "Magic Bullet: The History of Oral Rehydration Therapy." Medical History 38 (2006): 363–397.

Thomson, A. B. R., and E. A. Shaffer. "Physiology of the Colon," in *First Principles of Gastroenterology* (chapter 11). Gastroenterology Research Center. 1997. http://www.Gastroresource.com/en/ (accessed November 1, 2007).

Victora, C. G., J. Bryce, O. Fontaine, and R. Monasch. "Reducing Deaths from Diarrhoea through Oral Rehydration Therapy." *Bulletin of the World Health Organization* 78, no. 10, (2000).

Woo, D. D. F., T. Zeuthen, G. Chandy, and E. M. Wright. "Cotransport of Water by the Na+/glucose Cotransporter." *Proclamations of the National Academy of Science USA* 93 (1996): 13367–13370.

——. "The Management of Acute Diarrhea in Children: Oral Rehydration, Maintenance and Nutritional Therapy." MMWR. 1992.

——. "Cooking for the Gods: The Art of Home Ritual in Bengal." Mount Holyoke College Art Museum. 1998. http://www.mtholyoke.edu/offices/artmuseum/index.html (accessed September 27, 2007).

——. "The Oral Rehydration Therapy." Rainbow Pediatrics Knowledgebase. 2001. Rainbowpediatrics.net (accessed September 27, 2007).

——. "The History of Gatorade." 2003. Gatorade.com (accessed September 28, 2007).

——. "Sea Sick—Infection Outbreaks Challenge the Cruise Ship Experience." Water Quality and Health. 2004. http://www.waterandhealth.org/index.html (accessed November 9, 2007).

Chapter 5: Norman Borlaug:
A Green Revolution to Enhance Nutrition
Written by Billy Woodward and Debra Gordon

Bickel, Lennard. *Facing Starvation: Norman Borlaug and the Fight Against Hunger*. Pleasantville, NY: Reader's Digest Press, 1974.

Borlaug, Norman. "The Green Revolution, Peace, and Humanity." Nobel lecture. 1970. http://nobelprize.org/nobel_prizes/peace/laureates/1970/borlaug-lecture.html.

Borlaug, N. E. "Ending World Hunger: The Promise of Biotechnology and the Threat of Antiscience Zealotry." *Plant Physiology* 124 (2000): 487–490.

Brinkley, Douglas. "Bringing the Green Revolution to Africa: Jimmy Carter, Norman Borlaug, and the Global 2000 Campaign." *World Policy Journal* (March 22, 1996).

DeGregori, Thomas R. "Recognizing a Giant of Our Time: Dr. Borlaug Turns 90." American Council of Science and Health. 2004. http://www.acsh.org/news/newsid.625/News_detail.asp.

Doblhammer, G., and J. W. Vaupel. "Lifespan Depends on Month of Birth." *Proceedings of the National Academy of Sciences of the United States of America* 98, no.5, (2001): 2934–2939.

Easterbrook, Gregg. "Forgotten Benefactor of Humanity." *The Atlantic Monthly*, January 1997.

Fogel, Robert William. *The Escape from Hunger and Premature Death 1700–2000, Europe, America and the Third World*. New York: Cambridge University Press, 2004.

Hesser, Leon. *The Man Who Fed the World: Nobel Peach Prize Laureate Norman Borlaug and His Battle to End World Hunger*. Dallas, TX: Durban House, 2006.

Jain, Ajit. "Don't Be Afraid of New Technology." 2006. http://www.rediff.com//money/2006/sep/25mspec.html.

McCalla, A. F., and C. L. Revoredo. "Prospects for Global Feed Security: A Critical Appraisal of Past Projections and Predictions." International Food Policy Research Institute. 2001. http://www.ifpri.org/2020/dp/2020dp35.pdf (accessed September 14, 2006).

McFarland, Martha. "Sowing Seeds of Peace." 2003. Norman Borlaug Heritage Foundation. http://macserver.independence.k12.ia.us/~jlang/Education/BorlaugEssay.htm

Riley, James C. *Rising Life Expectancy: A Global History*. New York: Cambridge University Press, 2001.

Singh, Salil. "Norman Borlaug—A Billion Lives Saved." 2005. http://www.globalenvision.org/library/10/797.

Watters, Ethan. "DNA Is Not Destiny. *Discover Magazine*, November 2007.

"Undernourishment Around the World: Hunger and Mortality." The State of Food Security in the World. The Food and Nutrition Tehcnical Assistance Project. 2002. http://www.fantaproject.org/downloads/pdfs/FAOFS2002_6to13.pdf (accessed September 16, 2006).

Chapter 6: John Enders:
The Father of Modern Vaccines
Written by Billy Woodward and Joel Shurkin

Allen, A. *Vaccine—The Controversial Story of Medicine's Greatest Lifesaver.* New York: W. W. Norton, Company, 2007.

Bendiner, E. "Enders, Weller, and Robbins: The Trio That Fished in Troubled Waters." *Hospital Practice* 17, no.1, (January 1982): 163–197.

Chase, Allan. *Magic Shots.* New York: William Morrow & Company, Inc., 1982.

Driscoll, E. J., Jr. "John Enders." [The] *Boston Globe* (obituaries), September 10, 1985.

Eggars, H. J. "Milestones in Early Poliomyelitis Research: 1840–1949." *Journal of Virology* 73, no. 6, (1999): 4533–4535.

Enders, J. "A Note on Johnson's Staple of News." *Modern Language Notes* 40, no. 7, (1925).

——, T. H. Weller, and F. C. Robbins. "Cultivation of the Lansing Strain of Poliomyelitis Virus in Cultures of Various Human Embryonic Tissues." *Science* 109 (1949): 85–87.

——, F. C. Robbins, and T. H. Weller. "The Cultivation of Poliomyelitis Viruses in Tissue Culture." The Nobel Prize in Physiology or Medicine lecture. 1954. http://nobelprize.org/nobel_prizes/medicine/laureates/1954/enders-bio.html.

——, F. C. Robbins, and T. H. Weller. Nobel lecture. Nobelprize.org. 1954. http://nobelprize.org//nobel_prizes/medicine/laureates/1954/enders-robbins-wellerlecture.pdf (accessed March 29, 2007).

——, S. L. Katz, M. Milovanovic, and A. Holloway. "Studies on an Attenuated Measles-Virus Vaccine." *New England Journal of Medicine* 263, no. 4, (1960): 159–161.

——. "Vaccination Against Measles." Australian Journal of Experimental Biology 41 (1963): 467–490.

Feller, A. E., J. F. Enders, and T. H. Weller. "The Prolonged Coexistence of Vaccinia Virus in High Titre and Living Cells in Roller Tube Cultures of Chick Embryonic Tissues." *Journal of Experimental Medicine* 72 (1940): 367–388.

Galambos, L., and J. E. Sewell. *Networks of Innovation—Vaccine Development at Merck, Sharp & Dohme, and Mulford, 1985–1995*. Cambridge: Cambridge University Press, 1995.

"History of Vaccines." National Museum of American History. http://americanhistory.si.edu/polio/virusvaccine/history.htm (accessed March 27, 2007).

Ho, Monto. *Several Worlds: Reminiscences and Reflections of a Chinese-American Physician*. Singapore: World Scientific Publishing Company, 2005.

Hunt, R. "Vaccines—Past Successes and Future Prospects." Microbiology and Immunology On-line University of South Carolina. 2000–2006. http://pathmicro.med.sc.edu/lecture/vaccines.html (accessed March 29, 2007).

Kane, M., and H. Lasher. "The Case for Childhood Immunization." 2002. http://childrensvaccine.org/files/CVP_Occ_Paper5.pdf (accessed March 27, 2007).

Katz, Samuel L. "Elements of Style." *Harvard Medical Alumni Bulletin* (Winter, 1985): 20–23.

Katz, Samuel L. E-mail correspondence. March, 2008.

Kluger, Jeffrey. *Splendid Solution: Jonas Salk and the Conquest of Polio*. New York: G. P. Putnam's, 2004.

"Measles." Centers for Disease and Control. http://www.cdc.gov/vaccines/pubs/pinkbook/downloads/meas.pdf. Accessed February 26, 2007.

Moss, W., and D. Griffin. "Global Measles Elimination." Nature Reviews Microbiology 4 (December 2006): 900–908.

Orent, W. "Polio Is Almost Gone, But Will It Ever Be?" Massachusetts General Hospital. 2006. http://www.protomag.com/issues/2006_spring/polio_scourge_print.html (accessed March 30, 2007).

Oshinsky, D. M. *Polio: An American Story*. New York: Oxford University Press, 2005.

Paul, John. *Cell and Tissue Culture*. Baltimore: The Williams and Wilkins Company, 1970.

Podolsky, Lawrence M. Cures *Out of Chaos*. Amsterdam: Harwood Academic Publishers, 1998.

"Recommended Immunization Schedule for 0–6 Years." Centers for Disease Control and Prevention. http://www.cispimmunize.org/ IZSchedule_Childhood.pdf (accessed March 28, 2007).

Robbins, Alice. E-mail correspondence. March.

Robbins, Frederick C. "From Philology to the Laboratory." *Harvard Medical Alumni Bulletin* (Winter, 1985): 16–18.

Rosen, F. S. "Isolation of Poliovirus—John Enders and the Nobel Prize." *New England Journal of Medicine* 351, no. 15, (2004): 1481–1483.

Rosenberg, Nancy, and Louis Z. Cooper. *Vaccines and Viruses*. New York: Grosset & Dunlap, 1971.

Simmons, J. G. "John Franklin Enders: Persuading Viruses to Multiply," in *Doctors & Discoveries: Lives That Created Today's Medicine* (pp. 266–269). Boston: Houghton Mifflin, 2002.

Tyrrell, D. A. J. "John Franklin Enders." (Biographical Memoirs). *Biographical Memoirs of Fellows of the Royal Society* 33 (1987): 211–233.

Weller, Thomas H. *Growing Pathogens in Tissue Culture: Fifty Years in Academic Tropical Medicine, Pediatrics, and Virology.* Boston: Boston Medical Library, 2004.

Weller, T. H., and F. C. Robbins. "John Franklin Enders." *Biographical Memoirs of the National Academy of Sciences* 60 (1991):47–50.

——. "Vaccine Progress." *Time*, November 17, 1961.

——. "John Franklin Enders," in *Biographical Memoirs* (pp. 47–65). Washington, D.C.: National Academies Press, 1991.

——. "Achievements in Public Health, 1900–1999: Impact of Vaccines Universally Recommended for Children—United States, 1900–1998." *CDC Morbidity and Mortality Weekly Report* (April 2, 1999).

——. "Hundreds Gather to Remember Fred C. Robbins, MD." Case Western Reserve University Medical Bulletin 10, no. 1, (2004).

——. "Polio Vaccine: The Story Behind the Story." Children's Hospital Boston. 2005. http://www.childrenshospital.org/research/ Site2029/mainpagesS2029P6sublevel7Flevel9.html.

Chapter 7: Paul Müller:
DDT and the Prevention of Malaria
Written by Billy Woodward and Debra Gordon

Arguin, Paul, and Sonja Mali. "Travelers' Health," in *Yellow Book* (chapter 4: "Prevention of Specific Infectious Diseases). http://www.ias.ac.in/currsci/dec102003/1532.pdf (accessed July 20, 2007).

Carson, Rachel. *Silent Spring*. New York: First Mariner Books, 1962.

CDC—Malaria. "Protective Effect of Sickle Cell Trait Against Malaria—Associated Mortality and Morbidity. 2004. http://www.ced.gov/malaria/biology/sicklecell.html.

Desowitz, R. S. *Federal Body Snatchers and the New Guinea Virus: Tales of Parasites, People and Politics*. New York: W. W. Norton, 2002.

——. *The Malaria Capers: Tales of Parasites and People*. New York: W. W. Norton, 1991.

"DDT Ban Takes Effect." 1972. http://epa.gov/history/topics/ddt/01.html (accessed March 22, 2007).

Feldman, Stanley, and Vincent Marks. *Panic Nation*. London: John Blake Publishing, 2005.

Kwiatkowski, D. P. "How Malaria Has Affected the Human Genome and What Human Genetics Can Teach Us about Malaria." The American Journal of Human Genetics 77 (2005): 171–192.

"Malaria: Biology." CDC. 2004. http://www.cdc.gov/malaria/biology/index.html (accessed March 17, 2007).

"Malaria and the Red Cell." 2002. http://sickle.bwh.harvard.edu/malaria_sickle.html (accesseed April 2, 2007).

Minnesota Pollution Control Agency. "Creature of the Month—July." 2005. http://proteus.pca.state.mn.us/kids/c-july98.html.

McGrayne, Sharon Bertsch. *Prometheans in the Lab: Chemistry and the Making of the Modern World*. New York: McGraw-Hill, 2001.

Müller, P. Nobel Prize lecture. Nobel Prize.org. 1948. http://nobel-prize.org/Nobel_prizes/medicine/laureates/1948/muller-lecture.pdf.

Rocco, F. *The Miraculous Fever Tree: Malaria and the Quest for a Cure That Changes the World*. Great Britain: Harper Collins, 2003.

Tren, R., and R. Bate. *Malaria and the DDT Story*. London: Institute of Economic Affairs, 2001.

Saginaw County Mosquito Abatement Commission. 2007. http://www.scmac.org/Mosquito_trivia.html.

Sharma, V. P. "DDT: The Fallen Angel." *Current Science* 85, no. 11 (2003). http://www.ias.ac.in/currsci/dec102003/1532.pdf.

World Health Organization. "Malaria." 2007. http://www.who.int/topics/malaria/en/.

Chapter 8: Howard Florey:
Penicillin—The Miracle of Antibiotics
Written by Billy Woodward and Joel Shurkin

"Antimicrobial (Drug) Resistance." Exploring HIH. Ed. National Institute of Allergy and Infection. 2006. http://www.niaid.nih.gov/factsheets/antimicro.html.

Anderson, Dean W. *Praise the Lord and Pass the Penicillin*. Jefferson, NC: McFarland & Company, 2003.

Bickel, Lennard. *Florey, the Man Who Made Penicillin*. Melbourne: Sun Books, Pty Ltd., 1983.

Bloom, D. E, and K. Fang. "Social Technology and Human Health." University of Toronto-Scarborough. River Path Associates. 2001. http://www.utsc.utoronto.ca/~chan/

Istb01/readings/socialTecHuman-Health.pdf (accessed August 3, 2006).

Bowden, Mary Ellen. "Alexander Fleming: Pharmaceutical Achiever." Antibiotics in Action. 2002. http://www.chemheritage.org/educationservices/pharm/anitbiot/readings/Fleming.html.

Brown, Kevin. *Penicillin Man*. Gloucestershire: Sutton Books, 2004.

Bud, R. "Penicillin and the New Elizabethans." British Journal of Historical Science 31 (1998): 305–333.

Drexler, Madeline. *Secret Agents: The Menace of Emerging Infections*. Washington, D.C: Joseph Henry Press, 2002.

Fletcher, C. "First Clinical Use of Penicillin." *Br Med J* (Clin Res Ed) 289, no. 6460 (1984): 1721–1723.

Florey, H. W. "Penicillin." Nobel lecture. NobelPrize.org. 1945. http://nobelprize.org/Nobel_prizes/medicine/laureates/1945/florey-lecture.pdf.

Goldsworth, Peter D., and Alexander C. McFarlane. "Howard Florey, Alexander Fleming and the Fairy Tale of Penicillin." *Medical Journal of Australia* 176, no. 4, (2002): 176–178.

Hare, Ronald. "New Light on the History of Penicillin." *Medical History* 26, no. 1, (1982): 1–24.

Harris, Henry. "Howard Florey and the Development of Penicillin." *Notes and Records of the Royal Society* 53, no. 2, (1999): 243–252.

Hill, James. "James Hill's D-Day: 3rd Parachute Brigade." WW2 People's War. British Broadcasting Company. http://www.bbc.co.uk/ww-2peopleswar/stories/16/a2523016.shtml. Accessed February 16, 2007.

Lax, Eric. *The Mold in Dr. Florey's Coat.* New York: Henry Holt and Company, 2005.

Lewis, Ricki. "The Rise of Antibiotic-Resistant Infections." U.S. Food & Drug Administration. 1995. http://www.fda.gov/FDAC/features/795_antibio.html.

Macfarlane, Gwyn. *Howard Florey: The Making of a Great Scientist.* Oxford: Oxford University Press, 1979.

Master, David. *Miracle Drug: The Inner History of Penicillin.* London: Eyre & Spottiswoode, 1946.

Mines, Samuel. *Pfizer: An Informal History.* New York: Pfizer, 1978.

"Penicillin and Pharmacognosy." Drugstore Museum. Soderlund Village Drug. 2004. http://www.drugstoremuseum.com/sections/level_info2.php?level=1&level_id=196.

Saxon, Wolfgang. "Anne Miller, 90, First Patient Who Was Saved by Penicillin." [The] *New York Times*, June 9, 1999.

Sci/Tech. "Planet Bacteria." BBC News. 1998. http://news.bbc.co.uk/1/hi/sci/tech/158203.stm (accessed August 2, 2006).

Chapter 9: Frederick Banting:
Insulin—The First True Miracle Drug
Written by Billy Woodward and Debra Gordon

Bliss, M. *The Discovery of Insulin.* Chicago: University of Chicago Press, 1982.

——. "Resurrections in Toronto: The Emergence of Insulin." Horm Res 64, supplement 2, (2005): 98–102.

——. *Banting: A Biography*. Toronto: University of Toronto Press, 1984.

"Diabetes." World Health Organization. http://www.who.int/ mediacentre/factssheets/fs312/en/.

"Discovery and Early Development of Insulin, The." University of Toronto Libraries, Fisher Library Digital Collections. 2003. http://link.library.utoronto.ca/insulin/index.html.

Gale, E. A. "The Rise of Childhood Type 1 Diabetes in the 20th Century." *Diabetes* 51, no. 12, (2002): 3353–3361.

Gidney, R. D., and W. P. J. Millar. "Quantity and Quality: The Problem of Admissions in Medicine at the University of Toronto, 1910–1951." *Historical Studies in Education* 9, no. 2, (Fall, 1997).

Gillespie, K. M. "Type 1 Diabetes: Pathogenesis and Prevention." *CMAJ* 175, no. 2, (July 18, 2006): 165–170.

Herrington, G. M. "The Discovery of Insulin." *Ala Med* 64, no. 12, (June 1995): 6–12.

Hodges, B. "The Many and Conflicting Histories of Medical Education in Canada and the USA: An Introduction to the Paradigm Wars." *Medical Education* 39, no 6, (2005): 613–621.

Hume, S. E. *Frederick Banting: Hero, Healer, Artist*. Montreal: XYZ Publishing, 2001.

Jain, K. M., K. G. Swan, and K. F. Casey. "Nobel Prize Winners in Surgery." (Part 3—Frederick Grant Banting, Walter Rudolph Hess). *Am Surg* 48, no. 7, (July 1982): 287–290.

King, K. M., and Rubin, G. A. "A History of Diabetes: From Antiquity to Discovering Insulin." *British Journal of Nursing* 12, no. 18, (2003): 1091–1995.

Majumdar, S. "Glimpses of the History of Insulin." *Bull Ind Inst Hist Med* 31 (2001): 57–70.

"Novo Nordisk Is Changing Diabetes." Nov Nordisk. 2006. http:// www.novonordisk.com/about_us/facts_and_figures/facts.asp. (accessed July 3, 2006).

Rafter, G. W. "Banting and Best and the Sources of Useful Knowledge." *Perspect Biol Med* 26, no. 2, (Winter, 1983): 282–286.

Rosenfeld, L. "Insulin: Discovery and Controversy." *Clinical Chemistry* 48, no. 12, (2002): 2270–2288.

"Spark-Plug Man." *Time*, March 19, 1941. http://www.time.com/time/magazine/article/0,9171,765305,00.html.

Tattersall, R. B. "A Force of Magical Activity: The Introduction of Insulin Treatment in Britain, 1922–1926." *Diabet Med* 12, no. 9, (1995): 739–755.

Welbourn, R. "The Emergence of Endocrinology." *Gesnerus* 49 (1992): 137–150.

Chapter 10: Karl Landsteiner:
The Superman Scientist—Discoverer of Blood Groups
Written by Billy Woodward and Debra Gordon

America's Blood Centers. "56 Facts about Blood." 2007. http://www.americasbloood.org/go.cfm?do=Page.View&pid=12.

Baker, J. P., and S. L. Katz. "A History of Pediatric Specialties—Vaccines." Pediatric Research 55, no. 2, (2004): 347–356.

Bendiner, E. "Karl Landsteiner: Dissector of the Blood." *Hosp Pract* (Off Ed) 26, supplement 3A, (1991): 93–104.

Chudley, Albert E. "Genetic Landmarks through Philately—Karl Landsteiner: The Father of Blood Grouping." *Clinical Genetics* 57, no. 4, (2000): 267–269.

Dean, L. "Blood Groups and Red Cell Antigens." National Library of Medicine. http://www.ncbi.nlm.nih.gov/books/bv.fcgi?rid=rbcantigen.chapter.ch2 (accessed September 13, 2007).

Dodd, R. Y. "Bacterial Contamination and Transfusion Safety: Experience in the United States." *Transfusion clinique et biologique: journal de la Société française de transfusion sanguine* 10, no. 1, (2003): 6–9.

Eibl. M., W. R. Mayr, and G. J. Thorbeck (eds). *Epitope Recognition Since Landsteiner's Discovery*. New York: Springer-Verlag, 2002.

Figl, M., and L. E. Pelinka. "Karl Lansteiner: The Discover of Blood Groups." *Resuscitation* 63, no. 3, (2004): 251–254.

Fuller, E. "Karl Landsteiner." Innominate Society. 1973. http://www.innominatesociety.com/Articles/Karl%20Landsteiner.html (accessed September 12, 2007).

Giangrande, P. L. "The History of Blood Transfusion." *British Journal Haematology* 110, no. 4, (2000): 758–767.

Gottlieb, A. M. "Karl Landsteiner, the Melancholy Genius: His Time and His Colleagues, 1868–1943." *Transfusion Medicine Reviews* 12, no. 1, (1998): 18–27.

Hajdu. S. "Blood Transfusion from Antiquity to the Discovery of the Rh Factor." *Annals of Clinical and Laboratory Science* 33, no. 4, (2003): 471–473.

Heidelberger, M. "Karl Landsteiner." *Biographical Memoirs of the National Academy of Sciences* 40 (1969).

Hess, J. R., and P. J. Schmidt. "The First Blood Banker: Oswald Hope Robertson." *Transfusion* 40, no. 1, (2000): 110–113.

———. "Blood Use in War and Disaster: Lessons from the Past Centurey." *Transfusion* 43, no. 11, (2003): 1622–1633.

Hughes-Jones, N. C., and B. Gardner. "Red Cell Agglutination: The First Description by Creite (1869) and Further Observations Made by Landois (1875) and Landsteiner (1901)." *British Journal of Haematology* 119, no. 4, (2002): 889–893.

Kendrick, D. B. "Preservative Solutions." Blood Program in World War II. 1964. http://History.amedd.army.mil/booksdocs/wwii/blood/chapter9.html (accessed September 12, 2007).

Kirkman, E. "Blood Groups." *Anaethesia & Intensive Care Medicine* 8, no. 5, (2007): 200–202.

Learoyd, P. "A Short History of Blood Transfusion." National Blood Service. 2006.

http://hospital.blood.co.uk/library/pdf/training_education/history_of_transfusion.pdf (accessed September 12, 2007).

Landsteiner, K. "Ueber Agglutinationserscheinungen Normalen Menschlichen Blutes (On Agglutination Phenomena of Normal Human Blood)," in *Papers on Human Genetics*, S. H. Boyer (ed.) (pp.27–31). Englewood Cliffs, NJ: Prentice-Hall, 1901.

———. *The Specificity of Serological Reactions.* Mineola, NY: Dover Publications, 1962.

———. "On Individual Differences in Human Blood." Nobel Prize lecture. 1930. http://nobelprize.org/nobel_prizes/medicine/laureates/1930/landsteiner-lecture.html.

"Paul Ehrlich, Pharmaceutical Achiever." Chemical Heritage Foundation. 2001. http://www.chemheritage.org/EducationServices/pharm/chemo/readings/ehrlich.html (accessed September 11, 2007).

"Red Gold: The Epic Story of Blood." PBS. 2002. http://www.pbs. org/wnet/redgold (accessed September 14, 2007).

Rous, F. P. "Karl Landsteiner." *Fellows of the Royal Society* (Obiturary Notices) 5, no. 18, (1947): 295–324.

Schwarz, H. P., and F. Dorner. "Karl Landsteiner and His Major Contributions to Haematology." *British Journal Haematology* 121, no. 4, (2003): 556–565.

Speiser. P., and F. G. Smekal. *Karl Landsteiner*. Wien: Verlag Bruder Hollinek, 1975.

Starr, D. "Medicine, Money, and Myth: An Epic History of Blood." *Transfusion Medicine Reviews* 11, no 2, (2001): 119–121.

——. *Blood*. New York: Alfred A. Knopf, 2002.

Tagarelli, A., and et al. "Karl Landsteiner: A Hundred Years Later." *Transplantation* 72, no. 1, (2001): 3–7.

Tagliasacchi, D., and G. Carboni. "Let's Observe the Blood Cells." Fun Science Gallery. 2001. http://www.funsci.com/fun3_en/ blood/blood.html#3 (accessed September 3, 2007).

Wiener, A. S. "Karl Landsteiner, MD: History of Rh-Hr Blood Group System." *New York State Journal of Medicine* 69, no. 22, (1969): 2915–1935.

Young, J. H. "James Blundell (179–1878): Experimental Physiologist and Obstetrician." *Medical History* 8 (1964): 159–169.

Essay: A Choice for the Future:
Our Health or Our Wealth?
Written by Billy Woodward

"Bible Study Tools." Crosswalk.com 2007. http://bible.crosswalk. com/ (accessed January 11, 2008).

Cutler, David M. *Your Money or Your Life: Strong Medicine for America's Healthcare System*. New York: Oxford University Press, 2004.

"Health Insurance Coverage." National Coalition on Health Care. 2008. http://www.nchc.org/facts/coverage.shtml.

Kuhn, Deanna. "Connecting Scientific and Informal Reasoning." Merrill-Palmer Quarterly 39, no. 1, (1993): 74–103.

Manton, K. G., R. L. Lowrimore, A. D. Ullian, X. Gu, and H. D. Tolley. "Labor Force Participation and Human Capital Increases in an Aging Population and Implications for U.S. Research

Investment." *Proceedings of the National Academy of Sciences* 104, no. 26, (2007): 10802–10807.

Moses, H., III, E. R. Dorsey, D. H. M. Matheson, and S. O. Their. "Financial Anatomy of Biomedical Research." *Journal of the American Medical Association* 294, no. 11, (2005): 1333–1342.

Murphy, Kevin M., and Robert H. Topel. *Measuring the Gains from Medical Research: An Economic Approach.* Chicago: University of Chicago Press, 2003.

"Odds of Dying From...." National Safety Council. 2007. http://www.msc.org/research/odds.aspx (accessed January 12, 2008).

Wallenborn, White McKenzie. "George Washington's Terminal Illness: A Modern Medical Analysis of the Last Illness and Death of George Washington." The Papers of George Washington. 1999. http://gwpapers.virginia.edu/articles/wallenborn.html (accessed December 10, 2006).

Index